Lecture Notes in Physics

311

P. J. M. Bongaarts R. Martini (Eds.)

Complex Differential Geometry and Supermanifolds in Strings and Fields

Proceedings of the Seventh Scheveningen Conference, Scheveningen, The Netherlands, August 23–28, 1987

Springer-Verlag
Berlin Heidelberg GmbH

Editors

P. J. M. Bongaarts
Lorentz Institute, University of Leiden
NL-2311 SB Leiden, The Netherlands

R. Martini
Faculty of Applied Mathematics, University of Twente
P.O. Box 217, NL-7500 AE Enschede, The Netherlands

ISBN 978-3-662-13697-3 ISBN 978-3-540-45969-9 (eBook)
DOI 10.1007/978-3-540-45969-9

Originally published by Springer-Verlag Berlin Heidelberg New York in 1988
Softcover reprint of the hardcover 1st edition 1988

2158/3140-543210

PREFACE

The present volume contains the notes of lectures delivered at the seventh Scheveningen Conference held at Scheveningen, the Netherlands, August 23-28, 1987.

The conference was organized by P.J.M. Bongaarts (University of Leiden), E.M. de Jager (University of Amsterdam), H. Lemei (Delft University of Technology), R. Martini (University of Twente) and was financially supported by the Koninklijke Nederlandse Akademie van Wetenschappen.

This conference is the seventh in a series of meetings that started in 1973. All of these meetings have been centered around a few main topics of related interest and have been partly instructional in character, in order to stimulate further research. The last three conferences were devoted to aspects of differential geometry in mathematical physics, and this trend was continued in the present meeting. More specifically, the emphasis was this time on complex differential geometry (Kähler manifolds), super and graded manifolds, and their applications to supersymmetry, strings and field theory. The contributions in this volume can accordingly be divided roughly into:

1. Kähler manifolds in field theory: M. Forger, J. Wess.
2. Supersymmetry, superspace and graded manifolds: M. Batchelor, A. Rogers and P. Bryant, P.D. Jarvis.
3. String theory: D.G. Caldi, N. Sanchez.

The collection of lectures as a whole showed a remarkable unity of content and approach. The division into three subjects has therefore only an approximate value, as may be clear from the list of contents. For instance Kähler manifolds were also of importance in Caldi's lectures and "super" aspects, in those of Forger and Wess. Apart from the lectures mentioned in the list of contents there were also seminars by C. Peters and B. de Wit of which unfortunately no notes were available for this volume.

The organizing committee is indebted to all who made this conference possible and, in particular, to the Koninklijke Nederlandse Akademie van Wetenschappen for their financial support. We would like to express our gratitude to all who helped to make this conference a success, lecturers and participants alike.

Special thanks are due to E. Kamerich and J. Maassen from the university of Nijmegen who used their expert knowledge on computer programming and many hours of their time to solve for us a problem connected with the preparation of the typescript for this volume.
Finally, we wish to express our gratitude to Springer-Verlag for their assistance in the effective production of these proceedings.

Leiden, Enschede, June 1988
The Netherlands

P.J.M. Bongaarts
R. Martini

CONTENTS

M. Forger
Supersymmetric Sigma Models and Kähler Manifolds 1

J. Wess
Nonlinear Realizations, Kähler Manifolds and
the Virasoro Manifold 47

M. Batchelor
In Search of the Graded Manifold of Maps Between
Graded Manifolds 62

A. Rogers
Path Integration in Superspace 114

P. Bryant
Supermanifolds, Supersymmetry and Berezin Integration 150

P.D. Jarvis
Berezin Integration and Dimensional Reduction 168

D.G. Caldi
Aspects of String Theory 175

N. Sanchez
Field and String Quantization in Curved Space-Times 217

List of Participants 249

SUPERSYMMETRIC SIGMA MODELS AND KÄHLER MANIFOLDS

Michael Forger
Fakultät für Physik der Universität Freiburg,
Hermann-Herder-Str. 3, D-7800 Freiburg,
Fed. Rep. of Germany.

1. Introduction

The aim of these lectures is to discuss a number of issues that arise in connection with the theory of supersymmetric σ models and of Kähler manifolds - ranging from possible applications of supersymmetric σ models in particle physics to the explicit construction and classification of homogeneous Kähler manifolds. I shall begin by briefly reviewing the role of ordinary (=non-supersymmetric) σ models as effective low-energy theories, which appear whenever one is dealing with spontaneous symmetry breaking. This is of course an old and almost obsolete subject, but the supersymmetric extension of the picture has recently been revived. Namely, supersymmetric σ models have been proposed as effective low-energy theories for certain (as yet purely hypothetical) preon models - which implies, of course, that their physical predictions should be consistent with those of the standard model. There is no room here for describing these matters in detail, and I shall content myself with explaining why supersymmetry must play an essential role in such a scheme, and how concepts from the theory of complex manifolds and of Kähler manifolds enter the picture. I shall then turn to more mathematical aspects, and shall begin with the definition of Kähler manifolds and hyper-Kähler manifolds. The last section will then be devoted to a comprehensive discussion of homogeneous Kähler manifolds, including an explicit construction of Kähler potentials for arbitrary invariant Kähler metrics.

It is a pleasure to acknowledge the fruitful collaboration with M. Bordemann and H. Römer, on which the second part of these lecture notes is based. In addition, I want to thank both of them for reading the manuscript, and for various helpful suggestions, and W. Buchmüller for several enlightening discussions.

2. Sigma Models in Strong Interaction Physics

One of the main reasons why σ models have acquired such an important position in field theory is that they can serve as *effective field theories* at low energies[1] - irrespective of the (often complicated and also largely unknown) dynamical properties of the complete theory at high energies. The general philosophy is that particles which are unusually light (in an appropriate sense) should, in a first approximation, be considered as the massless *Goldstone particles* for some *spontaneously broken symmetry*; they may then acquire small masses due to a small additional explicit symmetry breaking, or even better, due to radiative corrections.

As an example, consider the original *σ model*, introduced in 1960 by Gell-Mann and Lévy [1] to describe the strong interactions between nucleons and pions. (See also [2,3] and [4], pp. 540-549.) The name "σ model" originates from one of their notations, namely the fact that in order to realize the desired chiral isospin symmetry (to be described below), they have to supplement the usual isodoublet Dirac field ψ for the nucleons and isotriplet pseudoscalar field $\vec{\pi}$ for the pions by an additional isoscalar scalar field, which they call σ.

More specifically, their starting point is the so-called *linear σ model*, for which they write down the total lagrangian

$$L \;=\; L_0 + L', \tag{2.1}$$

which is the sum of a "symmetric" term

$$\begin{aligned} L_0 \;=\; & \bar{\psi}\,[i\gamma\cdot\partial + g(\sigma+i\vec{\pi}\cdot\vec{\tau}\gamma_5)]\,\psi \\ & + \tfrac{1}{2}\,[(\partial\vec{\pi})^2 + (\partial\sigma)^2 - \mu^2(\vec{\pi}^2+\sigma^2)] - \tfrac{\lambda}{4}\,(\vec{\pi}^2+\sigma^2)^2 \end{aligned} \tag{2.2}$$

and a "symmetry breaking" term

$$L' \;=\; c\sigma. \tag{2.3}$$

Here, g and λ are coupling constants, μ is a mass parameter, the

---

1 This is the main application for σ models defined on four-dimensional space-time, but - as is of course well known - they are also extremely useful as two-dimensional toy models. As such, they form a subject in its own right, which I am not going to discuss in these notes.

τ's are the Pauli matrices (in isospin space), and c is a constant which signals the presence of some homogeneous external field that couples to σ. The symmetry alluded to above refers to the *chiral isospin group* SU(2)xSU(2), whose elements $(U,V) \in SU(2)xSU(2)$ act on the fields ψ, $\vec{\pi}$ and σ as follows:

$$\psi_+ \longrightarrow U\psi_+ \ , \quad \psi_- \longrightarrow V\psi_- \ , \tag{2.4}$$

$$(\sigma+i\vec{\pi}\cdot\vec{\tau}) \longrightarrow U\ (\sigma+i\vec{\pi}\cdot\vec{\tau})\ V^{-1}. \tag{2.5}$$

$$(\sigma-i\vec{\pi}\cdot\vec{\tau}) \longrightarrow V\ (\sigma-i\vec{\pi}\cdot\vec{\tau})\ U^{-1}. \tag{2.6}$$

Here,

$$\psi_\pm = \tfrac{1}{2}(1\pm\gamma_5)\psi \tag{2.7}$$

are the chiral components of ψ, and the transformation laws (2.5) and (2.6) (which are simply each other's inverse) have a natural interpretation in terms of the twofold covering homomorphism from SU(2)xSU(2) onto SO(4).

Namely, any complex (2x2)-matrix can be uniquely written in the form $A = \sigma+i\vec{\pi}\cdot\vec{\tau}$ with $\det A = \vec{\pi}^2+\sigma^2$, where $\vec{\pi}$ and σ can be arranged into a complex four-vector, and this four-vector will then be real if and only if A is special unitary up to a non-negative real scale factor, i.e. if and only if $A = \lambda U$ with $\lambda \geq 0$ and $U \in SU(2)$ (implying $\lambda = \sqrt{\det A}$). But this condition on A, as well as the value of the factor $\lambda = \sqrt{\det A}$, is invariant under multiplication of A by arbitrary matrices in SU(2), from the left or from the right, and so the prescription (2.5) defines an action of SU(2)xSU(2) on real four-vectors by matrices in SO(4). With a little bit of additional work, one can check that this construction does define a twofold covering homomorphism.

Now using the fact that

$$\bar{\psi}(\sigma+i\vec{\pi}\cdot\vec{\tau}\gamma_5)\psi = \bar{\psi}_+(\sigma+i\vec{\pi}\cdot\vec{\tau})\psi_- + \bar{\psi}_-(\sigma-i\vec{\pi}\cdot\vec{\tau})\psi_+,$$

one can check without any difficulty that L_0 is invariant under SU(2)xSU(2), while L' and hence L break the "axial part" ($V=U^{-1}$) of this symmetry and preserve only its "vector part" (V=U). Hence calculating the Noether currents corresponding to chiral isospin, one arrives at a conserved vector current (CVC)

$$\vec{V}_\mu = \tfrac{1}{2}\,\bar{\psi}\gamma_\mu\vec{\tau}\psi + \vec{\pi}\times\partial_\mu\vec{\pi} \quad , \quad \partial^\mu\vec{V}_\mu = 0 \quad , \tag{2.8}$$

and a partially conserved axial current (PCAC)

$$\vec{A}_\mu = \tfrac{1}{2}\,\bar{\psi}\gamma_\mu\gamma_5\vec{\tau}\psi + (\sigma\,\partial_\mu\vec{\pi} - \partial_\mu\sigma\,\vec{\pi}) \quad , \quad \partial^\mu\vec{A}_\mu = -\,c\vec{\pi} \quad . \tag{2.9}$$

In the quantum theory, the presence of the explicit symmetry breaking term L' will cause the field σ to develop a non-vanishing vacuum expectation value $v = \langle 0|\sigma|0\rangle$, so that in order to develop a sensible perturbation theory, one must replace it by the shifted field $\sigma'=\sigma-v$ and reexpress the lagrangian L in terms of ψ, $\vec{\pi}$ and σ' rather than ψ, $\vec{\pi}$ and σ. Dropping constants, one gets

$$\begin{aligned} L = {} & \bar{\psi}\,[i\gamma\cdot\partial + gv + g(\sigma'+i\vec{\pi}\cdot\vec{\tau}\gamma_5)]\,\psi \\ & + \frac{1}{2}\,[(\partial\vec{\pi})^2 - (\mu^2+\lambda v^2)\vec{\pi}^2 + (\partial\sigma')^2 - (\mu^2+3\lambda v^2)\sigma'^2] \\ & - \lambda v\sigma'(\vec{\pi}^2+\sigma'^2) - \frac{\lambda}{4}(\vec{\pi}^2+\sigma'^2)^2 + (c-\mu^2 v-\lambda v^3)\sigma' , \end{aligned} \tag{2.10}$$

showing that in the classical = tree approximation, the masses of the particles associated with the fields ψ, $\vec{\pi}$ and σ' are

$$m_\psi = -gv, \tag{2.11}$$

$$m_\pi = \sqrt{\mu^2+\lambda v^2}, \tag{2.12}$$

$$m_{\sigma'} = \sqrt{\mu^2+3\lambda v^2}. \tag{2.13}$$

Of particular interest is the limiting case c=0, in which the axial current is strictly conserved. Two essentially different situations may arise.

a) The *normal phase*, characterized by the condition that $\mu^2>0$. The minimum of the potential is located at $v^2=0$, the axial symmetry is fully restored, and $m_\psi=0$, $m_\pi=\mu$, $m_{\sigma'}=\mu$.

b) The *Goldstone phase*, characterized by the condition that $\mu^2<0$. The minimum of the potential is located at $v^2=-\mu^2/\lambda$, the axial symmetry is spontaneously broken, and $m_\psi=-gv$, $m_\pi=0$, $m_{\sigma'}=-\sqrt{2}\mu$.

It should be pointed out that the linear σ model is renormalizable (this follows by applying standard power counting arguments to the lagrangian L), and that the previous results on particle masses will remain qualitatively correct after renormalization - although the numerical relations between the various constants will change.

The main problem with the linear σ model is that it predicts a new particle σ', which is not observed in nature. It has therefore been proposed that this particle should have an extremely large mass, but unfortunately the model provides no explanation for why that should be so. The absence of such an explanation, or to put it more drastically, the apparent absence of the particle σ' as such, motivated Gell-Mann and Lévy to propose the so-called *non-linear σ model*, in which the bosonic fields are subjected to a quadratic constraint

$$\vec{\pi}^2+\sigma^2 = f_\pi^2 , \tag{2.14}$$

where f_π is a constant (the pion decay constant). The merit of this procedure is that it will, loosely speaking, eliminate one of the four bosonic fields, without spoiling chiral isospin invariance. One way to see that is to solve (2.14) for σ:

$$\sigma = (f_\pi^2-\vec{\pi}^2)^{\frac{1}{2}} . \tag{2.15}$$

Inserting this into (2.2) and (2.3), and dropping constants, one gets

$$\begin{aligned} L_0 &= \bar{\psi}\ [i\gamma\cdot\partial + g((f_\pi^2-\vec{\pi}^2)^{\frac{1}{2}}+i\vec{\pi}\cdot\vec{\tau}\gamma_5)]\ \psi \\ &\quad + \frac{1}{2}\ [\ (\partial\vec{\pi})^2 + \frac{(\vec{\pi}\cdot\partial\vec{\pi})^2}{(f_\pi^2-\vec{\pi}^2)}\ ] \end{aligned} \tag{2.16}$$

and

$$L' = c(f_\pi^2-\vec{\pi}^2)^{\frac{1}{2}} . \tag{2.17}$$

Thus σ becomes a composite field, and the problem with the missing particle disappears. In this form, the non-linear σ model provides a phenomenologically reasonable description of the strong interactions between nucleons and pions at sufficiently low energies - very much like the V-A four-fermion interaction model provides a phenomenologically reasonable description for the weak interactions at low energies. In fact, both models are non-renormalizable[2], and both models are (therefore) bound to fail at high energies: this happens when processes involving strange quarks resp. intermediate vector bosons become important.

---

[2] For example, the non-polynomial kind of interaction in (2.16) indicates that renormalizability of the original linear σ model has been lost.

The above formulation of the non-linear σ model is adapted to the spontaneous breakdown of (the "axial part" of) chiral isospin invariance, and the pions are simply the corresponding Goldstone bosons. It must however be pointed out that in the quantum theory, a constraint such as (2.14) cannot be imposed as an operator identity, and hence the status of what is meant by "eliminating the field σ" is not so clear. An alternative and maybe more rigorous approach is to return to the linear σ model in the Goldstone phase and take the limit $\lambda\to+\infty$, $\mu^2\to-\infty$ in such a way that $v^2=-\mu^2/\lambda=f_\pi^2$ remains constant. Indeed, inserting this condition into (2.2), and dropping constants, one arrives at the lagrangian

$$\begin{aligned} L_0 \;=\; & \bar{\psi}\,[i\gamma\cdot\partial + g(\sigma+i\vec{\pi}\cdot\vec{\tau}\gamma_5)]\,\psi \\ & + \frac{1}{2}\,[(\partial\vec{\pi})^2 + (\partial\sigma)^2] - \frac{\lambda}{4}\,(\vec{\pi}^2+\sigma^2-f_\pi^2)^2 \;, \end{aligned} \tag{2.18}$$

so that in the aforementioned limit, fluctuations of the bosonic fields which are transverse to the three-sphere defined by (2.14) will be suppressed. The same phenomenon will occur if one uses the lagrangian

$$\begin{aligned} L_0 \;=\; & \bar{\psi}\,[i\gamma\cdot\partial + g(\sigma+i\vec{\pi}\cdot\vec{\tau}\gamma_5)]\,\psi \\ & + \frac{1}{2}\,[(\partial\vec{\pi})^2 + (\partial\sigma)^2] - \lambda(\vec{\pi}^2+\sigma^2-f_\pi^2) \;, \end{aligned} \tag{2.19}$$

where λ is now a Lagrange multiplier field, rather than a coupling constant: this means that (2.19) is obtained from (2.18) upon replacing the potential term by a Lagrange multiplier term. On the other hand, (2.19) will yield (2.16) upon substituting the equation of motion for λ, which is exactly the constraint (2.14).

The general lesson to be drawn from the above example can be summarized as follows. Consider a field theory with a lagrangian L_0 which is invariant under some global internal symmetry group G, and suppose that the theory contains a multiplet of scalar fields which are subject to self-interactions with non-trivial vacuum structure, so that the G-symmetry is spontaneously broken. [Explicit symmetry breaking terms will from now on be disregarded.] More precisely, it is assumed that G is a compact connected Lie group, that the scalar fields in the theory can be arranged into a single scalar field with values in a given complex (or real) vector space E which carries a given unitary (or orthogonal) representation of G, that the lagrangian L_0 contains a potential V which is a G-invariant function on E, and that the set M of minima for V is a non-trivial submanifold of E ($M\neq\{0\}$). In such a situation, it is clear that -

at least for sufficiently low energies - fluctuations along M (massless modes) will predominate over fluctuations transversally to M (massive modes), and that it is therefore a reasonable low-energy approximation to replace the original E-valued scalar field by an M-valued scalar field φ which - apart from the property of being constrained to M[3] - is free (since V is constant along M). But this means that the dynamics of the field φ is given by the *non-linear σ model* on M, which is defined as follows.

The basic dynamical variable of the model is a scalar field $\varphi=\varphi(x)$ defined on ordinary space (usually Minkowski space-time or euclidean space) and taking values in a given *Riemann manifold* M. [Indefinite metrics (pseudo-riemannian manifolds) are excluded since they would violate the positive energy condition.] If U is a coordinate patch in M and (u^i) is a system of coordinates for M defined on U, then (at least over the domain $\varphi^{-1}[U]$) φ is represented by a set of ordinary scalar fields φ^i, and the dynamics of the model is given by the action

$$S = \frac{1}{2} \int d^4x \; g_{ij}(\varphi) \; \partial\varphi^i \; \partial\varphi^j \; , \tag{2.20}$$

where g is the *Riemann metric*, with components

$$g_{ij} = g(\partial/\partial u^i, \partial/\partial u^j).$$

By construction, the model is invariant under any Lie group G that acts on M by isometries.

In the spontaneous symmetry breaking situation discussed before, $M \subset E$ as a Riemann manifold (i.e., the metric on M is the one induced from the standard metric on E), and the field φ introduced above is called the *Goldstone field*. Moreover, M is automatically a disjoint union of G-orbits in E. The simplest situation is of course that where M is just a single G-orbit. Under this hypothesis, known as *absence of accidental degeneracies*, M becomes a *homogeneous Riemann manifold*, which means that G acts *transitively* on M by isometries. To make this more explicit, pick a(n arbitrary)

---

[3] Strictly speaking, this terminology is correct at the classical level only, because quantum fields - due to their short-distance singularities - cannot in general be subjected to algebraic constraints. Mathematically, this means that the constraint equations cannot be imposed as operator identities, and physically, it means that quantum fluctuations transverse to the submanifold cannot be completely suppressed. (The most notable exception is, of course, when the submanifold is a subspace and hence the constraint equation is linear.)

point m_0 in M, and denote its stability group in G by H. Moreover, let $\mathfrak{g}$ and $\mathfrak{h}$ be the Lie algebras of G and H, respectively, and choose an H-invariant subspace $\mathfrak{m}$ of $\mathfrak{g}$ which is a direct complement to the H-invariant subspace $\mathfrak{h}$ of $\mathfrak{g}$, i.e.,

$$\mathfrak{g} \;=\; \mathfrak{h} \oplus \mathfrak{m} \;. \tag{2.21}$$

Then one can identify

a) the manifold M with the homogeneous space G/H:

$$M \;\simeq\; G/H \;, \tag{2.22}$$

b) the tangent space $T_{m_0}M$ to M at m_0 with $\mathfrak{m}$:

$$T_{m_0}M \;\simeq\; \mathfrak{m} \;, \tag{2.23}$$

c) the metric tensor field g on M with an H-invariant inner product $(.,.)$ on $\mathfrak{m}$.

(The invariance statements refer to the action of H on $\mathfrak{g}$ obtained by restricting the adjoint representation of G on $\mathfrak{g}$ to H, and the H-invariant inner product on $\mathfrak{m}$ is normally obtained by restricting a G-invariant non-degenerate [but not necessarily positive definite] scalar product on $\mathfrak{g}$ to $\mathfrak{m}$. For a more detailed discussion of the related mathematics, see Sect. 4.1 of ref. [5].)

In physical terminology, it has become common practice to call the massless particles associated with the long-range quantum fluctuations of the M-valued Goldstone field φ *Goldstone bosons* or *quasi-Goldstone bosons*, according to whether they represent quantum fluctuations along the G-orbit G/H or transversal to the G-orbit G/H through the reference point m_0 in M. (Of course, such an interpretation is tacitly based on a perturbative approach, in which the constant field φ_0 that takes the value m_0 everywhere is used as a background field.) This implies that the Goldstone bosons form a multiplet under H corresponding to the aforementioned representation of H on $\mathfrak{m}$, while the quasi-Goldstone bosons are associated with accidental degeneracies.

3. Supersymmetric Sigma Models and Preon Theories

Two of the most important open problems in particle physics are the reason for the existence of several generations of quarks and leptons, and the origin of their masses. In the now generally accepted standard model for the weak, electromagnetic and strong interactions, for example, the number of generations is a free parameter[4], and the fermion masses are also undetermined, being functions of the Yukawa coupling constants which can be chosen at will [6]. Since grand unified theories are beset by the same kind of arbitrariness [6], it seems that the usual gauge theory approach is reaching its limits, and that new ideas are needed.

One of the most natural ideas in this direction is that quarks and leptons themselves are composite, i.e., are bound states of some more fundamental particles, which have collectively been named *preons*. It must be pointed out that this is pure speculation, since present experimental data give no evidence for compositeness: all one can say is that down to distances of $\sim 10^{-16}$cm, or equivalently, up to energies of ~200GeV, quarks and leptons behave as if they were point particles. Their masses, on the other hand, are very small when measured against this scale, with factors ranging from ~50 for the b-quark to ~400000 for the electron - not to speak of the neutrino(s). Hence if it is assumed that they are composite, one is immediately faced with a serious dynamical problem, namely: why are typical masses of bound states between preons so much smaller than their typical binding energy, i.e., the strength of interaction between them?

Hoping that the reader will, at this point, be struck by the "déjà vu" effect, I want to explain how the discussion in the last section can, in principle, be adapted to the present situation. The idea is, of course, that irrespective of the (unknown) dynamical properties of the underlying preon model, quarks and leptons should, in a first approximation, be the massless Goldstone particles for some spontaneously broken symmetry; their small masses may then be due to a small additional explicit symmetry breaking, or even better, to radiative corrections. The crucial difference to the previous situation is that quarks and leptons are *fermions* (spin ½) while pions are *bosons* (spin 0), and so *the Goldstone*

[4] According to present experimental evidence, there are (at least) three generations, while theory requires the number of generations to be ≤8 in order to maintain asymptotic freedom of QCD at short distances.

mechanism requires supersymmetry: in other words, the underlying preon model must be supersymmetric. There are then two different ways in which such a model can produce massless fermions.

a) Spontaneous breakdown of supersymmetry, leading to massless *Goldstone fermions*.

b) Spontaneous breakdown of an (ordinary) internal symmetry, leading to massless *Goldstone bosons*, plus massless *quasi-Goldstone bosons* and massless *quasi-Goldstone fermions*, which are simply their superpartners. Supersymmetry must then be broken at a lower scale, and in such a way that the Goldstone bosons and quasi-Goldstone bosons acquire masses, while the quasi-Goldstone fermions remain massless. This requires an additional mass protection mechanism, which is usually based on chirality.

Alternative b) has been proposed in 1983 by Buchmüller, Peccei and Yanagida [7] and has since then attracted a considerable amount of attention, from a phenomenological point of view as well as for purely theoretical reasons. (See, e.g., [8-15].) The theoretical interest is motivated by the fact that the effective field theory at low energies - or more precisely at energies below the internal symmetry breaking scale but above the supersymmetry breaking scale - should be a *supersymmetric σ model*.

The argument goes very much as in the ordinary (=non-supersymmetric) case, as follows. Consider a supersymmetric field theory with a lagrangian L_0 which is invariant not only under supersymmetry but also under some global internal symmetry group G, and suppose that the theory contains a multiplet of chiral superfields which are subject to self-interactions with non-trivial vacuum structure, so that the G-symmetry is spontaneously broken. [Explicit symmetry breaking terms will once again be disregarded.] More precisely, it is assumed that G is a compact connected Lie group, that the chiral superfields in the theory can be arranged into a single chiral superfield with values in a given complex vector space E which carries a given unitary representation of G, that the lagrangian L_0 contains a superpotential V which is a G-invariant function on E, and that the set M of minima for V is a non-trivial submanifold of E ($M \neq \{0\}$). Once again, it is clear that - at least for sufficiently low energies - fluctuations along M (massless modes) will predominate over fluctuations transversally to M (massive modes), and that it is therefore a reasonable low-energy approximation to replace the original E-valued chiral superfield by an M-valued chiral superfield Φ which - apart from the property of being constrained to M - is free (since V is constant along M). But this means that the dynamics of the field Φ is given by the *supersymmetric σ model* on M, which is defined as follows.

The basic dynamical variable of the model is a chiral superfield $\Phi = \Phi(x,\theta,\bar{\theta})$ defined on superspace and taking values in a given *Kähler manifold* M. [Indefinite metrics (pseudo-kählerian manifolds) are excluded, once again, since they would violate the positive energy condition.] If U is a coordinate patch in M and (z^α) is a system of complex coordinates for M defined on U, then (at least over the domain $\Phi^{-1}[U]$) Φ is represented by a set of ordinary chiral superfields Φ^α, and the dynamics of the model is given by the action

$$S = \int d^4x \; d^2\theta \; d^2\bar{\theta} \quad K(\Phi,\bar{\Phi}) \; , \tag{3.1}$$

where K is the *Kähler potential*. By construction, the model is invariant under any Lie group G that acts on M by biholomorphic isometries. Moreover, expanding Φ in terms of component fields such as

$$\begin{aligned} \varphi^\alpha(x) &= \Phi^\alpha(x,\theta=0,\bar{\theta}=0) \; , \\ \bar{\varphi}^\alpha(x) &= \bar{\Phi}^\alpha(x,\theta=0,\bar{\theta}=0) \; , \end{aligned} \tag{3.2}$$

and integrating over θ and $\bar{\theta}$, one arrives at the following expression for the purely scalar part of the action:

$$S = \int d^4x \quad g_{\alpha\bar{\beta}}(\varphi,\bar{\varphi}) \; \partial\varphi^\alpha \; \partial\bar{\varphi}^\beta \; , \tag{3.3}$$

where g is the *Kähler metric*, with components

$$g_{\alpha\bar{\beta}} = \frac{\partial^2 K}{\partial z^\alpha \; \partial \bar{z}^\beta} \; . \tag{3.4}$$

This is of course just the action for the ordinary non-linear σ model on M, written in complex coordinates.

In the spontaneous symmetry breaking situation discussed before, M⊂E as a Kähler manifold (i.e., the metric and the complex structure on M are the ones induced from the standard metric and the standard complex structure on E), and the field Φ introduced above is called the *Goldstone superfield*. Moreover, M is automatically a disjoint union of G-orbits in E, and is even a disjoint union of G^c-orbits in E, where G^c is the complex connected Lie group obtained from G, the given compact connected Lie group, by *complexification*: this is simply a consequence of the fact that the superpotential V must be a *holomorphic* function on E, so that

$$V \text{ G-invariant} \implies V \text{ } G^c\text{-invariant.}$$

The simplest situation is therefore that where M is just a single G^c-orbit. Under this hypothesis, M becomes a *homogeneous complex manifold*, but unfortunately M does not become a *homogeneous Kähler manifold*, simply because although G and G^c both act on M by biholomorphic transformations, the problem is that
a) G acts isometrically but not transitively.
b) G^c acts transitively but not isometrically.
Of course, M can be built up from its G-orbits, which are homogeneous Kähler manifolds and are therefore, as I shall explain in Sect. 5, completely classified. What remains is thus the problem of classifying all possible ways of how this can be done, and hence all possible choices for M itself. Some progress has recently been made in this matter [16], but a completely explicit answer has, to my knowledge, not been given so far.

It has been argued above that the supersymmetric extension of the non-linear σ model can be formulated if the target space for the fields is a Kähler manifold, but it is a fact - first observed by Zumino [17] - that conversely, the "only if" part in this statement is equally correct. More generally, it can be shown [17-20] that the possibility of defining a supersymmetric σ model with N-extended supersymmetry in d dimensions imposes the following conditions on the target manifold:

	d=2	d=4
N=1 supersymmetry	Riemann	Kähler
N=2 supersymmetry	Kähler	hyper-Kähler
N=4 supersymmetry	hyper-Kähler	

I shall not discuss the derivation of this result here, but I do want to explain, in the next section, the concepts of a Kähler manifold and of a hyper-Kähler manifold in somewhat greater detail.

4. Kähler Manifolds and Hyper-Kähler Manifolds

Very roughly speaking, Kähler/hyper-Kähler manifolds are the complex/hyper-complex analogue of Riemann manifolds. Therefore, it seems appropriate to explain first of all what complex and hyper-complex structures are.

There are essentially two approaches to the definition of a complex manifold, which may at first sight look rather different, but which turn out to be completely equivalent. The first method is to simply repeat the definition of a real manifold in terms of local coordinate systems and transition functions, with real vector spaces replaced by complex ones and real smooth functions or maps replaced by complex analytic = holomorphic ones; this is the point of view we have taken, e.g., in the preceding section. The second method starts out from a real manifold M, and defines an *almost complex structure* on M to be a tensor field I on M of type (1,1) satisfying

$$I^2 = -1 \ . \tag{4.1}$$

Such a structure is said to be *integrable*, or simply a *complex structure*, if in addition,

$$[IX,IY] - I[IX,Y] - I[X,IY] - [X,Y] = 0 \tag{4.2}$$

for all vector fields X,Y on M (where [.,.] denotes the Lie bracket of vector fields). For an interpretation of these conditions, note that the algebraic constraint (4.1) allows to view the tensor I_m, for every point m in M, as representing multiplication by i in the real tangent space T_mM to M at m, while the integrability condition (4.2) is imposed in order to guarantee that all these infinitesimal notions of multiplication by i fit together consistently as the base point varies over M.

In order to relate the two definitions, we shall find it convenient to consider a(n almost) complex structure I on M as being defined, by complex extension, not only on the real tangent bundle TM but also on the complexified tangent bundle T^cM. (Explicitly, this means that we shall write, for every point m in M, I_m not only for the given real linear transformation on T_mM but also for its unique extension to a complex linear transformation on T^c_mM.)

The main advantage in doing so is that on complex tangent vectors, I has eigenvalues ±i and can be diagonalized. In fact,

$$\begin{gathered} T^c M \;=\; T^{(1,0)}M \oplus T^{(0,1)}M \;, \\ \text{where} \\ I = \text{multiplication by } +i \text{ on } T^{(1,0)}M \;, \\ I = \text{multiplication by } -i \text{ on } T^{(0,1)}M \;, \end{gathered} \tag{4.3}$$

[Complex tangent vectors in $T^{(1,0)}M$ resp. $T^{(0,1)}M$ are called vectors of type (1,0) resp. (0,1).] Similarly, the tensor field I can be defined on the real cotangent bundle T^*M and on the complexified cotangent bundle $T^{*c}M=T^{c*}M$, and

$$\begin{gathered} T^{*c}M \;=\; T^{*(1,0)}M \oplus T^{*(0,1)}M \;, \\ \text{where} \\ I = \text{multiplication by } +i \text{ on } T^{*(1,0)}M \;, \\ I = \text{multiplication by } -i \text{ on } T^{*(0,1)}M \;. \end{gathered} \tag{4.4}$$

[Complex cotangent vectors in $T^{*(1,0)}M$ resp. $T^{*(0,1)}M$ are called covectors of type (1,0) resp. (0,1).] Now if M is a complex manifold and U is a coordinate patch in M, then any system (z^α) of complex coordinates for M defined on U will of course give rise to a system (x^α,y^α) of real coordinates for M defined on U, where

$$x^\alpha = \operatorname{Re} z^\alpha \quad , \quad y^\alpha = \operatorname{Im} z^\alpha \quad . \tag{4.5}$$

Thus M will carry a unique almost complex structure I such that

$$I(\partial/\partial x^\alpha) = +\,\partial/\partial y^\alpha \quad , \quad I(\partial/\partial y^\alpha) = -\,\partial/\partial x^\alpha \quad \text{on } TM|U \quad , \tag{4.6}$$

$$I(dx^\alpha) = +\,dy^\alpha \quad , \quad I(dy^\alpha) = -\,dx^\alpha \quad \text{on } T^*M|U \quad , \tag{4.7}$$

and

$$z^\alpha = x^\alpha + iy^\alpha \quad , \quad \bar{z}^\alpha = x^\alpha - iy^\alpha \quad \text{on } U \quad , \tag{4.8}$$

$$\begin{aligned} \partial/\partial z^\alpha &= \tfrac{1}{2}(\partial/\partial x^\alpha - i\,\partial/\partial y^\alpha) \quad \text{span} \quad T^{(1,0)}M|U \quad , \\ \partial/\partial \bar{z}^\alpha &= \tfrac{1}{2}(\partial/\partial x^\alpha + i\,\partial/\partial y^\alpha) \quad \text{span} \quad T^{(0,1)}M|U \quad . \end{aligned} \tag{4.9}$$

$$\begin{aligned} dz^\alpha &= dx^\alpha + i\,dy^\alpha \quad \text{span} \quad T^{*(1,0)}M|U \quad , \\ d\bar{z}^\alpha &= dx^\alpha - i\,dy^\alpha \quad \text{span} \quad T^{*(0,1)}M|U \quad . \end{aligned} \tag{4.10}$$

Moreover, it is easily checked that this almost complex structure is integrable, i.e., a complex structure. Conversely, we have

Theorem (*Newlander-Nirenberg*): Any complex structure on a real manifold turns it into a complex manifold.

Due to the theorem of Newlander-Nirenberg, we may think of a complex manifold M as being a real manifold equipped with a given complex structure. Similarly - and this is one of the advantages of such a point of view - we may think of a hyper-complex manifold M as being a real manifold equipped with a given hyper-complex structure, which is simply a collection of three complex structures satisfying the quaternion algebra. More precisely, given a real manifold M, one defines an *almost hyper-complex structure* on M to be a triple (I,J,K) of tensor fields I,J,K on M of type (1,1) satisfying

$$\begin{gathered} I^2 = -1 \ , \ J^2 = -1 \ , \ K^2 = -1 \ , \\ IJ = K = -JI \ , \ JK = I = -KJ \ , \ KI = J = -IK \ , \end{gathered} \tag{4.11}$$

and one says that such a structure is *integrable*, or simply a *hyper-complex• structure*, if each of these tensor fields is integrable in the sense of (4.2).

Next, let M be a real manifold which carries an almost complex structure I, and suppose t is a tensor field of type (0,2) on M. Then t is said to be *compatible* with I if

$$t(IX,IY) = t(X,Y) \tag{4.12}$$

for all vector fields X,Y on M. In particular, this definition applies to *pseudo-riemannian metrics* g as well as to *non-degenerate two-forms* ω. Moreover, setting

$$\omega(X,Y) = g(IX,Y) \ , \ g(X,Y) = \omega(X,IY) \tag{4.13}$$

establishes a one-to-one correspondence between compatible pseudo-riemannian metrics g on M and compatible non-degenerate two-forms ω on M, and whenever g and ω are related in this way, they are simply the real part and the imaginary part of a single *pseudo-hermitian metric* $\langle .,. \rangle$:

$$\langle X,Y \rangle = g(X,Y) + i\omega(X,Y) \ . \tag{4.14}$$

An especially important situation arises when the two-form ω is *closed*, and therefore defines a *symplectic structure*:

$$d\omega = 0 \ . \tag{4.15}$$

One then says that the corresponding metric is *pseudo-kählerian* and not just pseudo-riemannian.

Definition: A *Kähler manifold* is a real manifold carrying a complex structure I and a Riemann metric g which is compatible with the complex structure I, and is such that the corresponding two-form ω is closed. Such a Riemann metric g is called a *Kähler metric*, and the corresponding two-form ω is called the corresponding *Kähler form*.

Definition: A *hyper-Kähler manifold* is a real manifold carrying a hyper-complex structure (I,J,K) and a Riemann metric g which is compatible with each of the three complex structures I,J,K, and is such that each of the corresponding two-forms ω_I, ω_J, ω_K, given by

$$\begin{aligned} \omega_I(X,Y) &= g(IX,Y) \ , \\ \omega_J(X,Y) &= g(JX,Y) \ , \\ \omega_K(X,Y) &= g(KX,Y) \ , \end{aligned} \tag{4.16}$$

is closed. Such a Riemann metric is called a *hyper-Kähler metric*, and the corresponding two-forms ω_I, ω_J, ω_K are called the corresponding *hyper-Kähler forms*.

Returning to the situation where M is a real manifold which carries a complex structure I, assume that (z^α) is a system of complex coordinates in M defined on a coordinate patch U. Then for any compatible Riemann metric g on M,

$$g_{\alpha\beta} \equiv g(\partial/\partial z^\alpha,\partial/\partial z^\beta) = 0 \quad , \quad g_{\bar\alpha\bar\beta} \equiv g(\partial/\partial \bar z^\alpha,\partial/\partial \bar z^\beta) = 0 \quad , \tag{4.17}$$

and writing

$$g_{\alpha\bar\beta} \equiv g(\partial/\partial z^\alpha,\partial/\partial \bar z^\beta) \quad , \tag{4.18}$$

we find

$$ds^2 = g_{\alpha\bar\beta}\, dz^\alpha\, d\bar z^\beta \quad , \quad \omega = i\, g_{\alpha\bar\beta}\, dz^\alpha \wedge d\bar z^\beta \quad . \tag{4.19}$$

If in addition, g is a Kähler metric, then the coefficients $g_{\alpha\bar\beta}$ on U can be derived from a real-valued function K on U according to

$$g_{\alpha\bar\beta} = \frac{\partial^2 K}{\partial z^\alpha\, \partial \bar z^\beta} \quad . \tag{4.20}$$

Such a function K is called a *Kähler potential*; it is by no means unique but may be subjected to *gauge transformations* of the form

$$K(z,\bar{z}) \longrightarrow K(z,\bar{z}) + f(z) + \bar{f}(\bar{z}) \qquad (4.21)$$

with arbitrary holomorphic functions f of z and $\bar{f}$ of $\bar{z}$ (which need not be mutually conjugate): these will obviously leave the coefficients $g_{\alpha\bar{\beta}}$ of the Kähler metric g invariant. In particular, K will depend on the choice of coordinates, and under a change of coordinates, it does *not* transform like a tensor: so it behaves in many respects just like a gauge potential.

To conclude this section, I want to point out that there is a plethora of examples for Kähler manifolds, while examples for hyper-Kähler manifolds are somewhat less abundant, and seem harder to construct. For instance, there are two large classes of Kähler manifolds which can be considered as completely classified:

1. *Riemann surfaces:* These are precisely the complex one-dimensional Kähler manifolds. In fact, any surface M, that is, any oriented real two-dimensional manifold M, can be equipped with a complex structure I and a Kähler metric g, as follows:

   α) Construct a Riemann metric g on M, e.g., by using an appropriate partition of unity.

   β) Define ω to be the volume form on M induced by the Riemann metric g and the given orientation. Note that any two-form on M is automatically closed, since M has real dimension 2.

   γ) Define I to be the almost complex structure on M given by (4.13). (This means that one must check (4.1) by explicit computation.) Note that any almost complex structure on M is automatically integrable, since M has real dimension 2.

   Therefore, the classification problem is to determine, for a given surface M,

   a) the *moduli space* of M, i.e., the space of all complex structures I on M (up to equivalence),

   b) the space of all Kähler metrics g on M which are compatible with a given complex structure I on M (up to equivalence).

   [In view of the above construction, both of these spaces are necessarily non-empty.] I think it is fair to say that - at least for compact surfaces - this classification problem is completely solved, even though the geometric and analytic properties of moduli space (as opposed to Teichmüller space) are often difficult to handle. It should also be mentioned that the subject has recently played an important role in string theory, but there is no room here to discuss the matter in more detail, and I refer the interested reader to any one of the various reviews that have appeared recently (see, e.g., [21]).

2. *Homogeneous Kähler manifolds:* These are precisely the Kähler manifolds carrying a transitive proper action of a Lie group by biholomorphic isometries. (See Sect. 5.) Therefore, the classification problem is to determine, for a given homogeneous space M=G/H,

   a) the space of all G-invariant complex structures I on M (up to equivalence),

   b) the space of all G-invariant Kähler metrics g on M which are compatible with a given G-invariant complex structure I on M (up to equivalence).

   [In particular, not every homogeneous space M=G/H can be made into a homogeneous Kähler manifold, so it is part of the problem to find a necessary and sufficient condition for both of these spaces to be non-empty.] Once again, I think one can say that - at least for compact homogeneous spaces - this classification problem is completely solved, and I want to devote the rest of these lectures to explaining how that is done.

Besides Riemann surfaces and homogeneous Kähler manifolds, there exist other, more general types of Kähler manifolds. In algebraic geometry, for example, one studies complex algebraic varieties defined as (intersections of) sets of zeros of polynomials, which are of course Kähler manifolds provided that they are non-singular. The class of hyper-Kähler manifolds, on the other hand, seems to be much more restricted. Examples are real four-dimensional *Calabi-Yau spaces*, that is, real four-dimensional Riemann manifolds with zero Ricci tensor, *tangent bundles* or equivalently *cotangent bundles of hermitian symmetric spaces*, and appropriate neighbourhoods of the zero section in the tangent bundles or equivalently the cotangent bundles of homogeneous Kähler manifolds [24] (it is still an open question whether, in the last case, the construction of the hyper-Kähler metric can be extended from a neighbourhood of the zero section to the whole bundle). Further examples, and techniques for the construction of new hyper-Kähler metrics, can be found, e.g., in ref. [20].

5. Homogeneous Kähler Manifolds

It has already been mentioned in the previous section that a homogeneous Kähler manifold is, by definition, a Kähler manifold M carrying a transitive proper action of a Lie group G by biholomorphic isometries.[5] To make this more explicit, let us pick once and for all a(n arbitrary) reference point m_0 in M and write H for its stability group in G. Then M can be identified with the coset space G/H, and m_0 with the coset 1H=H of 1 in G; moreover, the resulting equality M=G/H will hold not only in the sense of sets (because the action is transitive) but also in the sense of manifolds (because the action is proper); see [23], pp. 264-266. In this sense, homogeneous manifolds and coset spaces are one and the same thing.

The subject that I want to discuss in this section is the classification problem for homogeneous Kähler manifolds M=G/H. To simplify matters, I shall not remain completely general, but shall impose a number of (rather mild) supplementary conditions.

1. Both M and G should be *connected*.
   Note that in general, homogeneity of M under G implies that all the connected components of M are identical copies of the same connected manifold M^0, and this will then be a connected Kähler manifold carrying a transitive proper action of a connected Lie group, namely the connected 1-component G^0 of G, by biholomorphic isometries; see [24], pp. 124-125.

2. The *redundancy group* of the action of G on M, i.e., the closed normal subgroup N of G consisting of those elements in G that act trivially on M, should be *finite*.
   A slightly stronger hypothesis would be that the action of G on M is *effective*, which means, by definition, that N is trivial. Note that in general, the action of G on M can always be made effective, since one may simply replace G by the quotient G/N, and H by the quotient H/N, if necessary (observe that N is obviously a closed normal subgroup not only of G but also of H). But I prefer to stick to the slightly weaker condition formulated above, since it is both strong enough to exclude unwanted

---

[5] The assumption that the action of G on M should be proper is of a technical nature: it is imposed to avoid topological complications which might otherwise occur when G is non-compact, but it is automatically satisfied when G is compact; see [23], p. 264. This explains why we have not encountered it before.

pathologies and general enough to cover the important examples, such as the action of $G = SU(N)$ on the generalized flag manifolds $G/H = SU(N)/S(U(N_1)x\ldots xU(N_p))$ $(N = N_1+\ldots+N_p)$, and in particular on the Grassmannians, which are precisely the generalized flag manifolds with p=2.

3. The group G is (i.e., can be chosen) *semisimple*.
   A somewhat weaker hypothesis would be that the group G is (i.e., can be chosen) *reductive*, which means, by definition, that the Lie algebra of G should be the direct sum of its centre with a semisimple ideal.

As we have seen, the first two of the three conditions above may always be imposed without any loss of generality, while the third one is truly restrictive, since there do of course exist homogeneous Kähler manifolds M=G/H for which G is not (i.e., cannot be chosen) semisimple. [The basic examples are the complex tori, for which the relevant group is abelian.] The restriction to semisimple groups is therefore not really a mathematical or physical necessity, but is made here mainly for mathematical convenience. In particular, it would be interesting to see to what extent, or rather with what modifications, the results presented below would continue to hold for reductive groups. Of course, such a generalization must unify the case where G is semisimple with the case where G is abelian, and it will provide a complete answer to the problem of classifying the *compact* homogeneous Kähler manifolds (simply because G can be chosen to be compact, and hence reductive, when M is compact). I am convinced that the necessary information can be found in the relevant mathematical literature, but have not investigated the matter any further.

With these preliminaries out of the way, I shall divide the material of this section into a number of steps.

Step 1: Reduce to a problem in Lie algebra theory.

Step 2: Determine the class of homogeneous spaces M=G/H for G that can be turned into homogeneous Kähler manifolds.

As it turns out, these homogeneous spaces are, at least in the compact case, precisely the adjoint orbits of G, i.e., the G-orbits under the adjoint representation Ad of the Lie group G on its Lie algebra $\mathfrak{g}$.

Step 3: Find the G-invariant complex structures I, the G-invariant Kähler metrics g and the G-invariant Kähler forms ω on the adjoint orbits of G.

Step 4: Determine explicit expressions for Kähler potentials.

Step 5: Find a pragmatic classification scheme.

Historically, it should be noted that the complete solutions to steps 1-3 have been known to mathematicians for a long time: they date back to the work of Borel [25], of Wang [26] and of Borel and Hirzebruch [27] in the 1950's. A detailed account, which I shall follow rather closely, is given in ref. [14]. The general solution to step 4, on the contrary, seems to have originated only recently: the main ideas are due to Bando, Kuramoto, Maskawa and Uehara [12] and to Itoh, Kugo and Kunitomo [13]. I shall present here a considerably simplified version of their construction, which has not been published before. Finally, an intuitively appealing as well as pragmatically useful classification scheme for homogeneous Kähler manifolds in terms of so-called *painted Dynkin diagrams* has been put forward in ref. [14]; it will only be described very briefly here.

I shall begin with steps 1 and 2, but let me first fix some notation and conventions. Throughout the following, M=G/H will be a connected homogeneous space, where G is a connected semisimple Lie group, and H is a closed subgroup, such that the redundancy group of the action of G on M is finite. The Lie algebras of G and of H will be denoted by $\mathfrak{g}$ and by $\mathfrak{h}$, respectively, and the tangent space $T_{m_0}M$ to M at the reference point m_0 will be identified with the orthogonal complement $\mathfrak{m}$ of $\mathfrak{h}$ in $\mathfrak{g}$ with respect to the *Killing form* of $\mathfrak{g}$ (which I shall always denote by Kill). This is legitimate due to the fact that the Killing form of $\mathfrak{g}$, as a G-invariant symmetric bilinear form on $\mathfrak{g}$, is non-degenerate on $\mathfrak{g}$ and negative definite on $\mathfrak{h}$, so that the two H-invariant subspaces $\mathfrak{h}$ and $\mathfrak{m}$ of $\mathfrak{g}$ have trivial intersection, and Kill is non-degenerate on $\mathfrak{m}$ as well. (See Sect. 2 of ref. [14] for an explicit proof.) Hence we arrive at an *H-invariant orthogonal direct decomposition*

$$\mathfrak{g} = \mathfrak{h} \oplus \mathfrak{m} \;, \tag{5.1}$$

and we shall write

$$X = X_{\mathfrak{h}} + X_{\mathfrak{m}} \tag{5.2}$$

for the decomposition of generators X in $\mathfrak{g}$ corresponding to (5.1). [The invariance statements refer to the adjoint representation Ad of the Lie group G on its Lie algebra $\mathfrak{g}$, and/or to the representations of the closed subgroup H on $\mathfrak{g}$, $\mathfrak{h}$ and $\mathfrak{m}$ obtained from it by restriction.] Thus H-invariance of the decomposition (5.1), i.e., of the two subspaces $\mathfrak{h}$ and $\mathfrak{m}$, implies the commutation relations

$$[\mathfrak{h},\mathfrak{h}] \subset \mathfrak{h} \quad , \quad [\mathfrak{h},\mathfrak{m}] \subset \mathfrak{m} \quad . \tag{5.3}$$

Note however that we do *not* in general require the commutation relation

$$[\mathfrak{m},\mathfrak{m}] \subset \mathfrak{h} \tag{5.4}$$

to hold as well, so that in the decomposition

$$[X,Y] = [X,Y]_{\mathfrak{h}} + [X,Y]_{\mathfrak{m}} \quad \text{for} \quad X,Y \in \mathfrak{m} \ , \tag{5.5}$$

neither of the two summands will vanish in general. In other words, M=G/H is a *reductive homogeneous space* but not necessarily a *symmetric space*.

Step 1: Reduce to a problem in Lie algebra theory.

The basic idea here is that the complex structure I, the Kähler metric g and the Kähler form ω on M=G/H, being G-invariant, must all be determined from their values I_0, g_0 and ω_0 at the reference point m_0=1H alone. As a result, it must be possible to translate essentially any set of conditions on the tensor fields I, g and ω over M, and in particular the one that makes up the definition of a Kähler manifold, into an analogous set of conditions on the tensors I_0, g_0 and ω_0 over $\mathfrak{m}$. Thus for example, G-invariance of I, g and ω (as tensor fields over M) will be equivalent to H-invariance of I_0, g_0 and ω_0 (as tensors over $\mathfrak{m}$).

To be more specific, note that the G-invariant complex structure I on M corresponds to an H-invariant linear transformation $I_0 : \mathfrak{m} \longrightarrow \mathfrak{m}$ on $\mathfrak{m}$ satisfying the analogue of (4.1),

$$I_0^2 = -1 \ . \tag{5.6}$$

Similarly, the G-invariant Kähler metric g on M and the G-invariant Kähler form ω on M correspond to an H-invariant scalar product (= positive definite symmetric bilinear form) g_0 on $\mathfrak{m}$

and to an H-invariant symplectic form (= non-degenerate antisymmetric bilinear form) ω_0 on $\mathfrak{m}$, respectively, satisfying the analogues of the compatibility condition (4.12),

$$\begin{aligned} g_0(I_0X,I_0Y) &= g_0(X,Y) \quad \text{for } X,Y\in\mathfrak{m} , \\ \omega_0(I_0X,I_0Y) &= \omega_0(X,Y) \quad \text{for } X,Y\in\mathfrak{m} , \end{aligned} \tag{5.7}$$

and the analogue of the relation (4.13),

$$\omega_0(X,Y) = g_0(I_0X,Y) \ , \ g_0(X,Y) = \omega_0(X,I_0Y) \quad \text{for } X,Y\in\mathfrak{m} . \tag{5.8}$$

Explicitly, H-invariance means that

$$I_0\mathrm{Ad}(h) = \mathrm{Ad}(h)I_0 \quad \text{for } h\in H . \tag{5.9}$$

$$g_0(\mathrm{Ad}(h)X,\mathrm{Ad}(h)Y) = g_0(X,Y) \quad \text{for } h\in H \ , \ X,Y\in\mathfrak{m} . \tag{5.10}$$

$$\omega_0(\mathrm{Ad}(h)X,\mathrm{Ad}(h)Y) = \omega_0(X,Y) \quad \text{for } h\in H \ , \ X,Y\in\mathfrak{m} . \tag{5.11}$$

Finally, with the notation introduced in (5.2) and (5.5), the integrability conditions (4.2) and (4.15) take the form

$$[I_0X,I_0Y]_{\mathfrak{m}} - I_0[I_0X,Y]_{\mathfrak{m}} - I_0[X,I_0Y]_{\mathfrak{m}} - [X,Y]_{\mathfrak{m}} = 0 \quad \text{for } X,Y\in\mathfrak{m} , \tag{5.12}$$

and

$$\omega_0([X,Y]_{\mathfrak{m}},Z) + \omega_0([Y,Z]_{\mathfrak{m}},X) + \omega_0([Z,X]_{\mathfrak{m}},Y) = 0 \quad \text{for } X,Y\in\mathfrak{m} , \tag{5.13}$$

respectively. For proofs see, e.g., [28], Vol. 2, pp. 200-201 & 219-221.

Step 2: Determine the class of homogeneous spaces M=G/H for G that can be turned into homogeneous Kähler manifolds.

As mentioned before, not every coset space M=G/H can be made into a homogeneous Kähler manifold. In fact, it is the existence of a G-invariant symplectic structure on M which imposes a non-trivial constraint.

To derive this constraint, we use the Killing form of $\mathfrak{g}$ to convert the H-invariant symplectic form ω_0 on $\mathfrak{m}$ to an H-invariant linear transformation $\Omega_0 : \mathfrak{m} \longrightarrow \mathfrak{m}$ on $\mathfrak{m}$, as follows:

$$\omega_0(X,Y) = \mathrm{Kill}(\Omega_0X,Y) \quad \text{for } X,Y\in\mathfrak{m} . \tag{5.14}$$

This is then extended to an H-invariant linear transformation $\Omega_0: \mathfrak{g} \longrightarrow \mathfrak{g}$ on $\mathfrak{g}$, as follows:

$$\begin{aligned} \Omega_0\big|_{\mathfrak{h}}: \mathfrak{h} \longrightarrow \mathfrak{h} \quad & \text{is zero.} \\ \Omega_0\big|_{\mathfrak{m}}: \mathfrak{m} \longrightarrow \mathfrak{m} \quad & \text{is given by (5.14).} \end{aligned} \tag{5.15}$$

The point is now that due to (5.13), Ω_0 must in fact be a *derivation* of $\mathfrak{g}$, which means that

$$\Omega_0[X,Y] = [\Omega_0 X,Y] + [X,\Omega_0 Y] \quad \text{for } X,Y\in\mathfrak{g} \ . \tag{5.16}$$

But $\mathfrak{g}$ is semisimple, so every derivation of $\mathfrak{g}$ is inner (see [24], p. 132). Hence there exists a generator Z^ω in $\mathfrak{g}$ such that

$$\Omega_0 X = [Z^\omega,X] \quad \text{for } X\in\mathfrak{g} \ . \tag{5.17}$$

Moreover, Ω_0 is zero on $\mathfrak{h}$ (by definition) and invertible on $\mathfrak{m}$ (since ω_0 on $\mathfrak{m}$ is non-degenerate), which implies that $\mathfrak{h}$ coincides with the centralizer of Z^ω in $\mathfrak{g}$, i.e.,

$$\mathfrak{h} = \{ X\in\mathfrak{g} \ / \ [X,Z^\omega]=0 \} \ . \tag{5.18}$$

In particular, Z^ω itself belongs to $\mathfrak{h}$. But in fact, much more than that is true. Namely, it can be shown that H is compact and connected, and that

$$H = \{ g\in G \ / \ \mathrm{Ad}(g)Z^\omega = Z^\omega \} \ . \tag{5.19}$$

(See Sect. 2 of ref. [14] for an explicit proof.) As a result, M can be identified with the adjoint orbit of G through Z^ω, and H can be viewed either as the stability group of the generator Z^ω in $\mathfrak{g}$ or as the centralizer of a torus $\tilde{T}$ in G.[6] Namely, define T^ω to be the closure, in G, of the one-parameter subgroup $\{\exp(tZ^\omega)/t\in\mathbb{R}\}$ of G generated by Z^ω, and $\hat{T}$ to be the centre of H. Then $T^\omega \subset \hat{T}$, and $\tilde{T}$ can be any torus in G such that $T^\omega \subset \tilde{T} \subset \hat{T}$. [Thus although H is uniquely determined by $\tilde{T}$, $\tilde{T}$ is not always uniquely determined by H.]

The results obtained so far are summarized in the following theorem, due to Borel [25].

---

6 A *torus* in G is, by definition, a compact connected abelian subgroup of G.

Theorem 1: *Let G be a connected semisimple Lie group and M be a Kähler manifold which is homogeneous under G, i.e., M carries a transitive proper action of G by biholomorphic isometries, with at most a finite redundancy group. Then M is an adjoint orbit of G. More precisely, M=G/H, where H, the stability group of some reference point* m_0 *in M, is a compact connected subgroup of G and is the stabilizer of a generator* Z_0 *in* $\mathfrak{g}$*, or equivalently, the centralizer of a torus* $\hat{T}$ *in G.*

What remains to be done, therefore, is to prove the converse statement, namely that adjoint orbits of semisimple Lie groups can, at least in the compact case, always be turned into homogeneous Kähler manifolds, and to enumerate all possible ways in which that can be done: this is the content of step 3.

Before passing on to the discussion of steps 3, 4 and 5, I want to impose one more supplementary condition, namely

4. Both M and G should be *compact*.

Once again, this condition is truly restrictive, since there do of course exist non-compact homogeneous Kähler manifolds. [The basic examples are the hermitian symmetric spaces of the non-compact type.] The restriction to compact groups and compact coset spaces is therefore, once again, not really a mathematical or physical necessity, but is made here in order to simplify the presentation. For a complete treatment of both the compact and non-compact cases (at least concerning step 3 and step 5), see ref. [14].

Next, let me fix some more notation and conventions, which will remain in force throughout the rest of these notes. In accordance with what has been said so far, M=G/H will be a compact connected homogeneous space, where G is a compact connected semisimple Lie group and H is the compact connected stability group of a given generator Z_0 in $\mathfrak{g}$ under the adjoint representation of G on $\mathfrak{g}$:

$$H = \{ g \in G \ / \ \mathrm{Ad}(g) Z_0 = Z_0 \} . \tag{5.20}$$

The centre of H will then be a torus $\hat{T}$ in G, and conversely, H will be the centralizer of $\hat{T}$ in G. Moreover, I shall choose, once and for all, a *maximal torus* T in G containing $\hat{T}$, so that

$$\hat{T} \subset T \subset H \subset G . \tag{5.21}$$

As usual, the corresponding Lie algebras will be denoted by the corresponding lower case german letters. In particular, $\mathfrak{h}$ is the stability algebra of Z_0 in $\mathfrak{g}$ under the adjoint representation ad of $\mathfrak{g}$ on $\mathfrak{g}$, i.e., the centralizer of Z_0 in $\mathfrak{g}$,

$$\mathfrak{h} = \{ X \in \mathfrak{g} \ / \ [Z_0, X] = 0 \} , \tag{5.22}$$

while $\hat{\mathfrak{t}}$ is the centre of $\mathfrak{h}$, and conversely, $\mathfrak{h}$ is the centralizer of $\hat{\mathfrak{t}}$ in $\mathfrak{g}$; moreover,

$$Z_0 \in \hat{\mathfrak{t}} \quad , \quad \hat{\mathfrak{t}} \subset \mathfrak{t} \subset \mathfrak{h} \subset \mathfrak{g} \ . \tag{5.23}$$

In addition, it is convenient to supplement the series of inclusions in (5.23) by introducing the appropriate orthogonal complements (here, as always, the notion of orthogonality refers to the Killing form of $\mathfrak{g}$). Namely, recall that $\mathfrak{m}$ is the orthogonal complement of $\mathfrak{h}$ in $\mathfrak{g}$: this defines the orthogonal direct decomposition (5.1), with commutation relations (5.3). Similarly, $\mathfrak{h}'$ and $\mathfrak{t}'$ will be the orthogonal complement of $\hat{\mathfrak{t}}$ in $\mathfrak{h}$ and in $\mathfrak{t}$, respectively: this defines orthogonal direct decompositions

$$\mathfrak{h} = \hat{\mathfrak{t}} \oplus \mathfrak{h}' , \tag{5.24}$$

$$\mathfrak{t} = \hat{\mathfrak{t}} \oplus \mathfrak{t}' , \tag{5.25}$$

with obvious commutation relations.

Now the Lie algebra $\mathfrak{g}$ and its complexification $\mathfrak{g}^c$ are semisimple (by hypothesis), and $\mathfrak{t}^c$ is a *Cartan subalgebra* of $\mathfrak{g}^c$. Hence if Δ is the corresponding *root system*, and the $\mathfrak{g}^{(\alpha)}$, $\alpha \in \Delta$, are the corresponding (one-dimensional) *root spaces*, given by

$$X_\alpha \in \mathfrak{g}^{(\alpha)} \iff [Z, X_\alpha] = \alpha(Z) X_\alpha \quad \text{for} \quad Z \in \mathfrak{t}^c , \tag{5.26}$$

we can write down the *root space decomposition* for $\mathfrak{g}^c$:

$$\mathfrak{g}^c = \mathfrak{t}^c \oplus \sum_{\alpha \in \Delta}^{\oplus} \mathfrak{g}^{(\alpha)} . \tag{5.27}$$

For later use, we introduce the generators $H_\alpha \in \mathfrak{t}^c$, $\alpha \in \Delta$, uniquely determined by the condition

$$\mathrm{Kill}(H_\alpha, H) = \alpha(H) \quad \text{for} \quad \alpha \in \Delta, \ H \in \mathfrak{t}^c , \tag{5.28}$$

as well as a set of generators $E_\alpha \in \mathfrak{g}^{(\alpha)}$, $\alpha \in \Delta$, subject to the normalization conditions

$$\mathrm{Kill}(E_\alpha, E_{-\alpha}) = 1 \quad \text{for} \quad \alpha \in \Delta \ , \tag{5.29}$$

and (hence) the commutation relations

$$[E_\alpha, E_{-\alpha}] = H_\alpha \quad \text{for} \quad \alpha \in \Delta \ . \tag{5.30}$$

In addition, we can and will require that

$$\overline{H}_\alpha = -H_\alpha = H_{-\alpha} \quad , \quad \overline{E}_\alpha = -E_{-\alpha} \quad \text{for} \quad \alpha \in \Delta \ , \tag{5.31}$$

where $\bar{.}$ denotes the conjugation in $\mathfrak{g}^c$ with respect to the (compact) real form $\mathfrak{g}$. (See Sect. 3 of ref. [14] for an explicit proof.) Moreover, we shall find it convenient to introduce a G-invariant inner product (= positive definite hermitian sesquilinear form) $\langle .,. \rangle$ on $\mathfrak{g}^c$ by setting

$$\langle X,Y \rangle = -\mathrm{Kill}(\overline{X},Y) \quad \text{for} \quad X,Y \in \mathfrak{g}^c \ , \tag{5.32}$$

and to denote hermitian adjoint (for linear transformations on $\mathfrak{g}^c$) with respect to this inner product by $.^+$, so that its G-invariance is expressed through the formula

$$\mathrm{Ad}(g)^+ = \mathrm{Ad}(g)^{-1} \quad \text{for} \quad g \in G \ , \tag{5.33}$$

implying, due to (5.31),

$$\mathrm{ad}(H_\alpha)^+ = \mathrm{ad}(H_\alpha) \quad , \quad \mathrm{ad}(E_\alpha)^+ = \mathrm{ad}(E_{-\alpha}) \quad \text{for} \quad \alpha \in \Delta \ , \tag{5.34}$$

which follows from combining (5.31) with the more general formula

$$\mathrm{ad}(X)^+ = -\mathrm{ad}(\overline{X}) \quad \text{for} \quad X \in \mathfrak{g}^c \ . \tag{5.35}$$

Finally, applying (5.26) with $Z_0 \in \mathfrak{t}^c$ shows that the complexified stability algebra $\mathfrak{h}^c$ is built up from the Cartan subalgebra $\mathfrak{t}^c$ and from those root spaces $\mathfrak{g}^{(\alpha)}$, $\alpha \in \Delta$, for which $\alpha(Z_0)=0$, while the complexified orthogonal complement $\mathfrak{m}^c$ of $\mathfrak{h}^c$ in $\mathfrak{g}^c$ is built up from those root spaces $\mathfrak{g}^{(\alpha)}$, $\alpha \in \Delta$, for which $\alpha(Z_0) \neq 0$. In other words, writing Δ as the disjoint union of two subsets Δ' and $\hat{\Delta}$ given by

$$\Delta' = \{ \ \alpha \in \Delta \ / \ \alpha(Z_0)=0 \ \} \tag{5.36}$$

$$\hat{\Delta} = \{ \ \alpha \in \Delta \ / \ \alpha(Z_0) \neq 0 \ \} \tag{5.37}$$

leads to

$$\mathfrak{h}^{c} = \mathfrak{t}^{c} \oplus \sum_{\alpha \in \Delta'}{}^{\oplus} \mathfrak{g}^{(\alpha)} , \qquad (5.38)$$

$$\mathfrak{m}^{c} = \sum_{\alpha \in \hat{\Delta}}{}^{\oplus} \mathfrak{g}^{(\alpha)} , \qquad (5.39)$$

which is why elements of Δ' and of $\hat{\Delta}$ will be called *subgroup roots* and *coset roots*, respectively. Note also that complexifying (5.1) and (5.3) gives rise to an orthogonal direct decomposition

$$\mathfrak{g}^{c} = \mathfrak{h}^{c} \oplus \mathfrak{m}^{c} , \qquad (5.40)$$

with commutation relations

$$[\mathfrak{h}^{c},\mathfrak{h}^{c}] \subset \mathfrak{h}^{c} \quad , \quad [\mathfrak{h}^{c},\mathfrak{m}^{c}] \subset \mathfrak{m}^{c} \quad , \qquad (5.41)$$

which, in view of (5.38), (5.39) and in accordance with (5.36), (5.37), are expressed through the following rules:

If the sum of two subgroup roots is a root,
it is a subgroup root:

$$\alpha,\beta \in \Delta' \; , \; \alpha+\beta \in \Delta \implies \alpha+\beta \in \Delta' . \qquad (5.42)$$

If the sum of a subgroup root and a coset root is a root,
it is a coset root:

$$\alpha \in \Delta' \; , \; \beta \in \hat{\Delta} \; , \; \alpha+\beta \in \Delta \implies \alpha+\beta \in \hat{\Delta} . \qquad (5.43)$$

Further insight into the significance of the decomposition (5.38) can be gained by noting that the stability algebra $\mathfrak{h}$ is not semisimple, but is compact and hence reductive, so it splits naturally, as a direct sum, into its centre $\hat{\mathfrak{t}}$ and its derived algebra $[\mathfrak{h},\mathfrak{h}]$, which is a semisimple ideal in $\mathfrak{h}$: of course, this decomposition is the one given in (5.24), and $\mathfrak{h}'=[\mathfrak{h},\mathfrak{h}]$ will be called the *semisimple part* of $\mathfrak{h}$. Thus, the Lie algebra $\mathfrak{h}'$ and its complexification $\mathfrak{h}'^{c}$ are semisimple (by construction), and $\mathfrak{t}'^{c}$ is a *Cartan subalgebra* of $\mathfrak{h}'^{c}$. Moreover, the corresponding *root system* can be identified with Δ', and the corresponding (one-dimensional) *root spaces* can be identified with the $\mathfrak{g}^{(\alpha)}$, $\alpha \in \Delta'$, which means that the *root space decomposition* for $\mathfrak{h}'^{c}$ is obtained by splitting off $\hat{\mathfrak{t}}^{c}$ from (5.38):

$$\mathfrak{h}'^{c} = \mathfrak{t}'^{c} \oplus \sum_{\alpha \in \Delta'}{}^{\oplus} \mathfrak{g}^{(\alpha)} , \qquad (5.44)$$

Step3a: Find the G-invariant complex structures I on M=G/H.

According to step 1, and after complexification (as explained in Sect. 4), we have to look for all complex linear transformations $I_0: \mathfrak{m}^c \longrightarrow \mathfrak{m}^c$ on $\mathfrak{m}^c$ which are real, i.e., which commute with the conjugation $\bar{\cdot}$ in $\mathfrak{g}^c$ with respect to the (compact) real form $\mathfrak{g}$, and which satisfy (5.6), (5.9) (H-invariance) and (5.12) (integrability). But using the arguments given so far, one can show that all such linear transformations must have the form

$$I_0 E_\alpha = i\varepsilon_\alpha E_\alpha \quad \text{with } \varepsilon_\alpha = \pm 1 \quad \text{for} \quad \alpha \in \hat{\Delta} \ , \tag{5.45}$$

where the signs ε_α, $\alpha \in \hat{\Delta}$, satisfy the reality condition

$$\varepsilon_{-\alpha} = -\varepsilon_\alpha \quad \text{for} \quad \alpha \in \hat{\Delta} \ , \tag{5.46}$$

the H-invariance condition

$$\varepsilon_{\alpha+\beta} = \varepsilon_\beta \quad \text{for} \quad \alpha \in \Delta', \ \beta \in \hat{\Delta} \quad \text{such that} \quad \alpha+\beta \in \hat{\Delta} \ , \tag{5.47}$$

and the integrability condition

$$\begin{gathered} \varepsilon_\alpha = \varepsilon_\beta \implies \varepsilon_\alpha = \varepsilon_{\alpha+\beta} = \varepsilon_\beta \\ \text{for} \quad \alpha,\beta \in \hat{\Delta} \quad \text{such that} \quad \alpha+\beta \in \hat{\Delta} \ . \end{gathered} \tag{5.48}$$

Such a collection of signs amounts to an *invariant ordering* in $\hat{\Delta}$, that is, to a decomposition $\hat{\Delta} = \hat{\Delta}^+ \cup \hat{\Delta}^-$ such that $\hat{\Delta}^- = -\hat{\Delta}^+$ and

$$\alpha \in \Delta' \ , \ \beta \in \hat{\Delta}^\pm \ , \ \alpha+\beta \in \hat{\Delta} \implies \alpha+\beta \in \hat{\Delta}^\pm \ , \tag{5.49}$$

$$\alpha,\beta \in \hat{\Delta}^\pm \ , \ \alpha+\beta \in \hat{\Delta} \implies \alpha+\beta \in \hat{\Delta}^\pm \ , \tag{5.50}$$

defined by putting

$$\alpha \in \hat{\Delta}^\pm \iff \alpha \in \hat{\Delta} \ , \ \varepsilon_\alpha = \pm 1 \ . \tag{5.51}$$

Besides the notion of an invariant ordering in $\hat{\Delta}$, there are however other, more geometric ways to encode the information contained in the choice of an invariant complex structure on M=G/H. One of them involves what I shall call a *canonical basis* $\hat{B}$ of $\hat{\mathfrak{t}}$, and the other is in terms of a *Weyl chamber* $\hat{C}$ in $\hat{\mathfrak{t}}$. (The notion of a Weyl chamber in $\hat{\mathfrak{t}}$ is defined in complete analogy with that of a Weyl chamber in the usual sense: it is one of the connected components of the subset of $\hat{\mathfrak{t}}$ which arises by removing the union of all the hyperplanes in $\hat{\mathfrak{t}}$ on which the coset roots α in $\hat{\Delta}$ annihilate.)

To describe how an invariant ordering in $\hat{\Delta}$ gives rise to the desired canonical basis $\hat{B}$ as well as to the Weyl chamber $\hat{C}$, we proceed as follows. First of all, choose an auxiliary ordering in Δ': this simply serves to extend the given invariant ordering in $\hat{\Delta}$ to an ordering in the entire root system Δ. Write $B(\Delta)$ for the ***basis** of **simple roots*** and $B(\Lambda)$ for the ***basis of fundamental weights*** (basis of the weight lattice Λ) corresponding to this latter ordering: these two bases are related to each other by $B(\Delta)=\{\alpha_1,\ldots,\alpha_r\}$, $B(\Lambda)=\{\lambda_1,\ldots,\lambda_r\}$, with $r=\dim\mathfrak{t}$, and

$$2\,\frac{(\lambda_j,\alpha_k)}{(\alpha_k,\alpha_k)} \equiv 2\,\frac{\mathrm{Kill}(H_{\lambda_j},H_{\alpha_k})}{\mathrm{Kill}(H_{\alpha_k},H_{\alpha_k})} = \delta_{jk} \quad \text{for } 1\le j,k\le r\ . \tag{5.52}$$

Obviously, we can split $B(\Delta)$ and $B(\Lambda)$ into disjoint subsets

$$B(\Delta) = B'(\Delta) \cup \hat{B}(\Delta)\quad,\quad B'(\Delta)=\{\alpha'_1,\ldots,\alpha'_{r'}\}\ ,\ \hat{B}(\Delta)=\{\hat{\alpha}_1,\ldots,\hat{\alpha}_{\hat{r}}\}\ ,$$
$$B(\Lambda) = B'(\Lambda) \cup \hat{B}(\Lambda)\quad,\quad B'(\Lambda)=\{\lambda'_1,\ldots,\lambda'_{r'}\}\ ,\ \hat{B}(\Lambda)=\{\hat{\lambda}_1,\ldots,\hat{\lambda}_{\hat{r}}\}\ ,$$

with $r'=\dim\mathfrak{t}'$ and $\hat{r}=\dim\hat{\mathfrak{t}}$, defined by putting $B'(\Delta)=B(\Delta)\cap\Delta'$, $\hat{B}(\Delta)=B(\Delta)\cap\hat{\Delta}$. Then $B'(\Delta)$ is a basis of the root subsystem Δ', and by construction, the set

$$\hat{B} = \{\ -iH_{\hat{\lambda}_1}\ ,\ \ldots\ ,\ -iH_{\hat{\lambda}_{\hat{r}}}\ \} \tag{5.53}$$

consists of vectors belonging to $\hat{\mathfrak{t}}$ (because they belong to $\mathfrak{t}$ and are annihilated by the simple subgroup roots in $B'(\Delta)$ and hence by all the subgroup roots in Δ'), and since these vectors are also linearly independent, they do constitute a basis of $\hat{\mathfrak{t}}$. The Weyl chamber $\hat{C}$ in $\hat{\mathfrak{t}}$ is then the set of all linear combinations of vectors in $\hat{B}$ with strictly positive coefficients:

$$\hat{C} = \{\ -i\sum_{j=1}^{\hat{r}} c_j H_{\hat{\lambda}_j} \in \hat{\mathfrak{t}}\ /\ c_j>0 \ \text{ for } 1\le j\le\hat{r}\ \}\ . \tag{5.54}$$

The important point is now that although $B(\Delta)$ and $B'(\Delta)$ and $\hat{B}(\Delta)$, as well as $B(\Lambda)$ and $B'(\Lambda)$, all depend on the choice of the auxiliary ordering in Δ', $\hat{B}(\Lambda)$, $\hat{B}$ and $\hat{C}$ do not: they are fully determined by the given invariant ordering in $\hat{\Delta}$ alone. [In fact, any other choice for the auxiliary ordering in Δ' can be related to the original one by a Weyl group transformation w which maps the corresponding bases $B_1(\Delta)$ and $B_2(\Delta)$ of simple roots into each other and is the identity on $\hat{t}$, which implies directly that $\hat{B}_1$ and $\hat{B}_2$ will coincide.] In particular, this justifies the term "canonical basis" for $\hat{B}$.

Conversely, it is clear that a Weyl chamber $\hat{C}$ determines a unique invariant ordering in $\hat{\Delta}$, since not only

$$\begin{aligned} \hat{C} &= \{ Z\in\hat{t} \ / \ i\alpha(Z)>0 \quad \text{for } \alpha\in\hat{\Delta}^+ \} , \\ \hat{C} &= \{ Z\in\hat{t} \ / \ i\alpha(Z)<0 \quad \text{for } \alpha\in\hat{\Delta}^- \} , \end{aligned} \tag{5.55}$$

but also

$$\begin{aligned} \hat{\Delta}^+ &= \{ \alpha\in\hat{\Delta} \ / \ i\alpha(Z)>0 \quad \text{for } Z\in\hat{C} \} . \\ \hat{\Delta}^- &= \{ \alpha\in\hat{\Delta} \ / \ i\alpha(Z)<0 \quad \text{for } Z\in\hat{C} \} . \end{aligned} \tag{5.56}$$

In conclusion, note that, as in (4.3),

$$\mathfrak{m}^c = \mathfrak{m}^{(1,0)} \oplus \mathfrak{m}^{(0,1)} , \tag{5.57}$$

with

$$\mathfrak{m}^{(1,0)} = \sum_{\alpha\in\hat{\Delta}^+}{}^{\oplus} \mathfrak{g}^{(\alpha)} , \quad \mathfrak{m}^{(0,1)} = \sum_{\alpha\in\hat{\Delta}^-}{}^{\oplus} \mathfrak{g}^{(\alpha)} , \tag{5.58}$$

and that due to (5.38)-(5.43), (5.49), (5.50), (5.57), (5.58),

$$\hat{\mathfrak{h}} = \mathfrak{h}^c \oplus \mathfrak{m}^{(0,1)} . \tag{5.59}$$

is a Lie subalgebra of $\mathfrak{g}^c$, such that

$$\mathfrak{g}^c = \hat{\mathfrak{h}} \oplus \mathfrak{m}^{(1,0)} . \tag{5.60}$$

In terms of the root space decomposition for $\mathfrak{g}^c$,

$$\hat{\mathfrak{h}} = \mathfrak{t}^c \oplus \sum_{\alpha\in\Delta'\cup\hat{\Delta}^-}{}^{\oplus} \mathfrak{g}^{(\alpha)} . \tag{5.61}$$

Step 3b: Find the G-invariant Kähler metrics g and the G-invariant Kähler forms ω on M=G/H.

According to step 1, and after complexification, we have to look for all symmetric complex bilinear forms g_0 and antisymmetric complex bilinear forms ω_0 on $\mathfrak{m}^c$ which are real, i.e., which intertwine between the conjugation $\bar{\cdot}$ in $\mathfrak{g}^c$ with respect to the (compact) real form $\mathfrak{g}$ and ordinary complex conjugation in $\mathbb{C}$, and which satisfy (5.7) (compatibility), (5.8), (5.10) & (5.11) (H-invariance) and (5.13) (integrability). But using the arguments given so far, one can show that all such bilinear forms must be determined from a single generator Z in $\hat{\mathfrak{t}}$, as follows:

$$g_0(E_\alpha, E_\beta) = 0 \quad \text{for} \quad \alpha, \beta \in \hat{\Delta} \quad \text{with} \quad \alpha+\beta \neq 0 \ ,$$
$$\omega_0(E_\alpha, E_\beta) = 0 \quad \text{for} \quad \alpha, \beta \in \hat{\Delta} \quad \text{with} \quad \alpha+\beta \neq 0 \ , \tag{5.62}$$

$$g_0(E_\alpha, E_{-\alpha}) = -i\epsilon_\alpha \alpha(Z) \quad \text{for} \quad \alpha \in \hat{\Delta} \ .$$
$$\omega_0(E_\alpha, E_{-\alpha}) = \alpha(Z) \quad \text{for} \quad \alpha \in \hat{\Delta} \ . \tag{5.63}$$

Moreover, g_0 and ω_0 will be non-degenerate if and only if

$$\alpha(Z) \neq 0 \quad \text{for} \quad \alpha \in \hat{\Delta} \ , \tag{5.64}$$

and g_0 will be positive definite if and only if Z belongs to the Weyl chamber $\hat{C}$ in $\hat{\mathfrak{t}}$ that was introduced above to represent the given complex structure on M (see (5.55), (5.56)).

These results are summarized in the following theorem; a more detailed proof, and the extension to the non-compact case, are given in ref. [14].

Theorem 2: *Let G be a compact connected semisimple Lie group and M be an adjoint orbit of G, i.e., M=G/H, where H, the stability group of some reference point m_0 in M, is a compact connected subgroup of G and is the stabilizer of a generator Z_0 in $\mathfrak{g}$, or equivalently, the centralizer of a torus $\tilde{T}$ in G. Then M can be turned into a Kähler manifold which is homogeneous under G. More precisely, the center of H is a torus $\hat{T}$ in G, and we have the following statements.*

1. *The set of all G-invariant complex structures I on M is in one-to-one correspondence with the set of all Weyl chambers $\hat{C}$ in the center $\hat{\mathfrak{t}}$ of $\mathfrak{h}$: this is a finite set.*
2. *Fixing a G-invariant complex structure I on M with corresponding Weyl chamber $\hat{C}$, the set of all G-invariant Kähler metrics g and Kähler forms ω on M is in one-to-one correspondence with the points in the Weyl chamber $\hat{C}$: this is a convex cone in $\hat{\mathfrak{t}}$ which has a canonical basis $\hat{B}$ consisting of vectors in $\hat{\mathfrak{t}}$ which are dual to $\hat{r}$ of the fundamental weights $\lambda_1, \ldots, \lambda_r$.*

Throughout the preceding discussion, the class of compact homogeneous Kähler manifolds has been analyzed in terms of compact coset space representations, that is, coset space representations $M=G/H$ in which *all* spaces involved are compact. In other words, G and H are compact Lie groups, with H a closed subgroup of G: it is then obvious that M must be compact, but the price to be paid is that G and H fail to be complex, so the quotient can only be taken in the real sense, and the complex structure on M has to be given separately. There is however an alternative approach where the situation is essentially reversed; here, the class of compact homogeneous Kähler manifolds is analyzed in terms of complex coset space representations, that is, coset space representations $M=\tilde{G}/\tilde{H}$ in which *all* spaces involved are complex. In other words, $\tilde{G}$ and $\tilde{H}$ are complex Lie groups, with $\tilde{H}$ a closed subgroup of $\tilde{G}$; the quotient can and should then be taken in the complex sense, so the complex structure on M is built in automatically, but the price to be paid is that $\tilde{G}$ and $\tilde{H}$ fail to be compact, so it is far from obvious that M must be compact. There are thus (at least) two rather different useful ways of writing the same manifold M as a coset space, but it is clear that they must be closely related, and that it must in fact be possible to determine $\tilde{G}$ and $\tilde{H}$, say, when G and H are given. (Conversely, it must be equally possible to determine G and H when $\tilde{G}$ and $\tilde{H}$ are given, but this question will not concern us here.) More concretely, suppose that G and H are matrix groups: this can be achieved without loss of generality, and it allows to define the complexifications G^c of G and H^c of H.[7] Then $\tilde{G}$ equals G^c, but $\tilde{H}$ is strictly larger than H^c: it is the connected complex Lie subgroup $\hat{H}$ of G^c generated by the complex Lie subalgebra $\hat{\mathfrak{h}}$ of $\mathfrak{g}^c$ defined in (5.59). To show that this is indeed the correct choice, one must show that the map from G/H to $G^c/\hat{H}$ induced by the embedding of G into G^c is a diffeomorphism, and this is obvious at least at the infinitesimal level since

$$T_{1H}(G/H) \simeq \mathfrak{m} \quad \text{(cf. (5.1))} \qquad \text{and} \qquad T_{1\hat{H}}(G^c/\hat{H}) \simeq \mathfrak{m}^{(1,0)} \quad \text{(cf. (5.60))}$$

are isomorphic as vector spaces. In fact, using the exponential map $\exp: \mathfrak{g}^c \longrightarrow G^c$, one can introduce a particularly nice system of complex coordinates z^α in M, defined on an appropriate open neighbourhood of the reference point $m_0=1\hat{H}$ in $M=G^c/\hat{H}$, by setting

7 Recall that the notion of complexification is well defined for Lie algebras, while at the group level, it makes sense for matrix groups, but not for Lie groups in general. (A famous example of a real Lie group which has no complexification is the universal covering group of $SL(2,\mathbb{R})$.) On the other hand, M is an adjoint orbit, so one can always replace G and H by the matrix groups Ad(G) and Ad(H) obtained as their images under the adjoint representation of G on $\mathfrak{g}$, if necessary.

$$m(z) = g(z)\hat{H} \in M \quad , \quad g(z) = \exp\Big[\sum_{\alpha\in\hat{\Delta}^+} z^\alpha E_\alpha \Big] \in G^c \; . \qquad (5.65)$$

Note also that fixing the complex subgroup $\hat{H}$ of G^c amounts to fixing not only the compact subgroup H of G, but the complex structure I on the coset space G/H (specified, e.g., by the direct decomposition (5.59)) as well. The complex subgroups $\hat{H}$ of G^c resp. complex subalgebras $\hat{\mathfrak{h}}$ of $\mathfrak{g}^c$ obtained in this way are precisely the *parabolic subgroups* of G^c resp. *parabolic subalgebras* of $\mathfrak{g}^c$, which have been studied extensively in the mathematical literature; the reader is referred to [29], Chapters 8-10, for an account from the point of view of linear algebraic groups.

One of the situations where the complex coset space representation becomes useful is in the construction of Kähler potentials for the G-invariant Kähler metrics given in Theorem 2 above.

Step 4: Determine explicit expressions for Kähler potentials.

According to ref. [12], candidates for Kähler potentials $K(\eta)$ can be defined in terms of appropriate projection operators η onto $\hat{H}$-invariant subspaces of representation spaces V for G^c. More precisely, suppose that

a) V is a given finite-dimensional complex vector space carrying a given inner product (= positive definite hermitian sesquilinear form) $\langle .,. \rangle$.

b) π is a given analytic representation of G^c or $\mathfrak{g}^c$ on V whose restriction to the compact real form, as a representation of G or $\mathfrak{g}$ on V, is unitary, i.e.,

$$\pi(g)^+ = \pi(g)^{-1} \quad \text{for} \quad g\in G \; , \qquad (5.66)$$

$$\pi(X)^+ = -\pi(X) \quad \text{for} \quad X\in \mathfrak{g} \; , \qquad (5.67)$$

and more generally,

$$\pi(X)^+ = -\pi(\bar{X}) \quad \text{for} \quad X\in \mathfrak{g}^c \; , \qquad (5.68)$$

where $.^+$ denotes hermitian adjoint (for linear transformations on V) with respect to the inner product $\langle .,. \rangle$, and $\bar{.}$ denotes the conjugation in $\mathfrak{g}^c$ with respect to the (compact) real form $\mathfrak{g}$. [It should be noted that the representation π of G^c on V and the representation π of $\mathfrak{g}^c$ on V are conceptually different objects, but for the sake of simplicity, I shall use the same letter π to denote them both, since it will always be clear from the context which one is meant, and also since in most of the concrete examples, both of them are, in terms of matrices, given by the same explicit formula.]

c) η is the orthogonal projector onto an $\hat{H}$-invariant subspace $V(\eta)$ of V, i.e., a linear transformation $\eta: V \longrightarrow V$ on V satisfying

$$\eta^2 = \eta \quad , \quad \eta^+ = \eta \quad , \quad \eta\, \pi(\hat{h})\, \eta = \pi(\hat{h})\, \eta \quad \text{for} \quad \hat{h}\in\hat{H} \; .$$
$$V(\eta) = \mathrm{im}(\eta) \; . \tag{5.69}$$

Given η, and given a linear transformation $S: V \longrightarrow V$ on V,

$$\det_\eta(S) \;=\; \det(1-\eta+\eta S\eta) \; . \tag{5.70}$$

will denote the determinant of its diagonal block that maps $V(\eta)$ to itself and is the identity on $V(\eta)^\perp$.

Note that according to (5.32)-(5.35), a natural choice for V and π is to take $V=\mathfrak{g}^c$ and to let π be the adjoint representation.

Assuming that V, π and η are given, the construction of the Kähler potential $K(\eta)$ proceeds as follows. First, recall that the natural projection from G^c onto $M=G^c/\hat{H}$ defines a holomorphic principal $\hat{H}$-bundle. Thus we can pick a holomorphic local section σ of this bundle, defined on an open subset U of M, say, and put

$$K_\eta^\sigma(m) \;=\; \ln \det_\eta\big(\pi(\sigma(m))^+\; \pi(\sigma(m))\big) \quad \text{for} \quad m\in U \; . \tag{5.71}$$

Of course, we can equally well pick any other holomorphic local section τ of this bundle, defined on an open subset V of M, say, and put

$$K_\eta^\tau(m) \;=\; \ln \det_\eta\big(\pi(\tau(m))^+\; \pi(\tau(m))\big) \quad \text{for} \quad m\in V \; . \tag{5.72}$$

Then on the overlap $U\cap V$, τ is related to σ by a holomorphic transition function $\hat{h}: U\cap V \longrightarrow \hat{H}$ according to

$$\tau(m) \;=\; \sigma(m)\; \hat{h}(m) \quad \text{for} \quad m\in U\cap V \tag{5.73a}$$

and it follows from (5.69), (5.70) that

$$K_\eta^\tau(m) - K_\eta^\sigma(m) \;=\; \ln \det_\eta(\pi(\hat{h}(m))) + \ln \det_\eta(\pi(\hat{h}(m))^+)$$
$$\text{for} \quad m\in U\cap V \; . \tag{5.74a}$$

More generally, if $g \in G$, then on the overlap $U \cap g^{-1} \cdot V$, the left translate of τ by g^{-1} is related to σ by a holomorphic transition function $\hat{h}(g,.): U \cap g^{-1} \cdot V \longrightarrow \hat{H}$, which in addition depends smoothly on g, according to

$$\tau(g \cdot m) = g\, \sigma(m)\, \hat{h}(g,m) \quad \text{for } m \in U \cap g^{-1} \cdot V , \tag{5.73b}$$

and it follows from (5.66), (5.69), (5.70) that

$$K_\eta^\tau(g \cdot m) - K_\eta^\sigma(m) = \ln \det_\eta(\pi(\hat{h}(g,m))) + \ln \det_\eta(\pi(\hat{h}(g,m))^+) \quad \text{for } m \in U \cap g^{-1} \cdot V . \tag{5.74b}$$

But this is precisely the transformation law required from a (collection of) Kähler potential(s) that will give rise to a G-invariant Kähler metric; see (4.20), (4.21). It is therefore sufficient to describe the Kähler potential in terms of a specific holomorphic local section σ and a specific system of complex local coordinates z^α, and the standard choice made by the physicists is the one given in (5.65); thus

$$\begin{aligned} K(\eta)(z,\bar{z}) &= \ln \det_\eta\left(\pi(g(z))^+ \, \pi(g(z))\right) , \\ g(z) &= \exp\left[\sum_{\alpha \in \hat{\Delta}^+} z^\alpha E_\alpha \right] , \end{aligned} \tag{5.75}$$

and

$$g(\eta)_{\alpha\bar{\beta}}\,(z,\bar{z}) = \frac{\partial^2 K(\eta)(z,\bar{z})}{\partial z^\alpha\, \partial \bar{z}^\beta} . \tag{5.76}$$

By construction, $g(\eta)$ is G-invariant, so it suffices to calculate $g(\eta)$ at $z=0$, $\bar{z}=0$. This can be done by using the formula

$$\frac{d}{dt} \ln \det(A(t)) = \operatorname{trace}\left[A(t)^{-1} \frac{d}{dt} A(t) \right] ,$$

which implies

$$\begin{aligned} &\frac{\partial^2}{\partial s \partial t} \ln \det(A(s,t)) \\ &= \operatorname{trace}\left[A(s,t)^{-1} \frac{\partial^2}{\partial s \partial t} A(s,t) \right. \\ &\qquad \left. - A(s,t)^{-1} \frac{\partial}{\partial s} A(s,t)\, A(s,t)^{-1} \frac{\partial}{\partial t} A(s,t) \right] . \end{aligned}$$

The result of this short calculation is, due to (5.68)-(5.70) and (5.31),

$$g(\eta)_{\alpha\bar{\beta}}\ (z=0,\bar{z}=0) \;=\; \mathrm{trace}\Big[\ \eta\ \pi(E_{-\beta})\ (1-\eta)\ \pi(E_\alpha)\ \eta\ \Big]\ . \qquad (5.77)$$

This expression is still quite cumbersome and does not really provide any insight, but it can be simplified dramatically by making systematic use of (5.69).[8] Namely, consider the infinitesimal version of the invariance condition in (5.69), together with its adjoint:

$$\begin{aligned} \eta\ \pi(X)\ \eta &= \pi(X)\ \eta \quad \text{for } X\epsilon\hat{\mathfrak{h}}\ . \\ \eta\ \pi(X)^+\ \eta &= \eta\ \pi(X)^+ \quad \text{for } X\epsilon\hat{\mathfrak{h}}\ . \end{aligned} \qquad (5.78)$$

But since $\mathfrak{h}^c$ is invariant under the conjugation $\bar{\cdot}$ in $\mathfrak{g}^c$ with respect to the (compact) real form $\mathfrak{g}$, (5.68) shows that this formula can be sharpened as follows:

$$[\pi(Z),\eta] = 0 \quad \text{for } Z\epsilon\, \mathfrak{t}^c\ . \qquad (5.79)$$

$$[\pi(E_\alpha),\eta] = 0 \quad \text{for } \alpha\epsilon\Delta'\ . \qquad (5.80)$$

$$\begin{aligned} \eta\ \pi(E_\alpha)\ \eta &= \eta\ \pi(E_\alpha) \quad \text{for } \alpha\epsilon\hat{\Delta}^+\ . \\ \eta\ \pi(E_\alpha)\ \eta &= \pi(E_\alpha)\ \eta \quad \text{for } \alpha\epsilon\hat{\Delta}^-\ . \end{aligned} \qquad (5.81)$$

Now assume that $\alpha,\beta\epsilon\hat{\Delta}^+$ with $\alpha\neq\beta$. Then there exists $Z\epsilon\,\mathfrak{t}^c$ such that $\alpha(Z)\neq\beta(Z)$, and using

$$[\pi(Z),\pi(E_\gamma)] = \pi([Z,E_\gamma]) = \gamma(Z)\pi(E_\gamma)\ ,$$

with $\gamma=\alpha$ or $\gamma=-\beta$, together with $\eta^2=\eta$ and cyclicity of the trace, we obtain

$$\begin{aligned} &(\alpha(Z)-\beta(Z))\ \mathrm{trace}\Big[\ \eta\ \pi(E_{-\beta})\ (1-\eta)\ \pi(E_\alpha)\ \eta\ \Big] \\ &\quad= \mathrm{trace}\Big[\ \eta\ \pi(E_{-\beta})\ (1-\eta)\ [\pi(Z),\pi(E_\alpha)]\ \eta\ \Big] \\ &\qquad+ \mathrm{trace}\Big[\ \eta\ [\pi(Z),\pi(E_{-\beta})]\ (1-\eta)\ \pi(E_\alpha)\ \eta\ \Big] \\ &\quad= \mathrm{trace}\Big[\ \eta\ \pi(E_{-\beta})\ (1-\eta)\ [\pi(Z),\pi(E_\alpha)]\ \Big] \\ &\qquad+ \mathrm{trace}\Big[\ \eta\ [\pi(Z),\pi(E_{-\beta})]\ (1-\eta)\ \pi(E_\alpha)\ \Big] \\ &\quad= \mathrm{trace}\Big[\ \eta\ \pi(E_{-\beta})\ [\pi(Z),(1-\eta)\ \pi(E_\alpha)]\ \Big] \\ &\qquad+ \mathrm{trace}\Big[\ [\pi(Z),\eta\ \pi(E_{-\beta})]\ (1-\eta)\ \pi(E_\alpha)\ \Big] \qquad \text{by (5.79)} \\ &\quad= \mathrm{trace}\Big[\ [\pi(Z),\eta\ \pi(E_{-\beta})\ (1-\eta)\ \pi(E_\alpha)]\ \Big] \\ &\quad= 0\ . \end{aligned}$$

---

[8] I am indebted to M. Bordemann for suggesting the idea of the following simple argument (my original proof of eq. (5.83) below was much more complicated).

The same kind of reasoning shows that for $\alpha \epsilon \hat{\Delta}^+$,

$$\text{trace}\Big[\ \eta\ \pi(E_{-\alpha})\ (1-\eta)\ \pi(E_\alpha)\ \eta\ \Big]$$
$$= \text{trace}\Big[\ \eta\ \pi(E_{-\alpha})\ \pi(E_\alpha)\ \eta\ \Big] - \text{trace}\Big[\ \eta\ \pi(E_{-\alpha})\ \eta\ \eta\ \pi(E_\alpha)\ \eta\ \Big]$$
$$= \text{trace}\Big[\ \eta\ \pi(E_{-\alpha})\ \pi(E_\alpha)\ \eta\ \Big] - \text{trace}\Big[\ \pi(E_{-\alpha})\ \eta\ \eta\ \pi(E_\alpha)\ \Big] \qquad \text{by (5.81)}$$
$$= -\ \text{trace}\Big[\ \eta\ [\pi(E_\alpha),\pi(E_{-\alpha})]\ \eta\ \Big]$$
$$= -\ \text{trace}\Big[\ \eta\ \pi([E_\alpha,E_{-\alpha}])\ \Big]$$
$$= -\ \text{trace}\Big[\ \eta\ \pi(H_\alpha)\ \Big]\ ,$$

while for $\alpha \epsilon \Delta'$,

$$0 = \text{trace}\Big[\ \eta\ \pi(E_{-\alpha})\ \pi(E_\alpha)\ \eta\ \Big] - \text{trace}\Big[\ \pi(E_{-\alpha})\ \eta\ \eta\ \pi(E_\alpha)\ \Big] \qquad \text{by (5.80)}$$
$$= -\ \text{trace}\Big[\ \eta\ [\pi(E_\alpha),\pi(E_{-\alpha})]\ \eta\ \Big]$$
$$= -\ \text{trace}\Big[\ \eta\ \pi([E_\alpha,E_{-\alpha}])\ \Big]$$
$$= -\ \text{trace}\Big[\ \eta\ \pi(H_\alpha)\ \Big]\ .$$

Therefore, there exists a generator $Z(\eta)$ in $\hat{\mathfrak{t}}$ which is uniquely determined by the condition that

$$\text{trace}\Big[\ \eta\ \pi(H_\alpha)\ \Big] = -i\alpha(Z(\eta)) \quad \text{for} \quad \alpha \epsilon \Delta\ , \tag{5.82}$$

and we see that, in accordance with (5.62), (5.63),

$$g(\eta)_{\alpha\bar{\beta}}\ (z=0,\bar{z}=0) = i\delta_{\alpha\beta}\alpha(Z(\eta)) \quad \text{for} \quad \alpha,\beta \epsilon \hat{\Delta}^+\ . \tag{5.83}$$

The only question that remains is whether $g(\eta)$ will be positive definite. As we shall see, this is usually not the case, but there is a class of examples for which $g(\eta)$ is at least positive semidefinite, and which is sufficiently general to yield Kähler potentials, for all the G-invariant Kähler metrics given in Theorem 2 before, by taking linear combinations with strictly positive coefficients.

The examples in question are all based on taking $V=\mathfrak{g}^c$ and letting π be the adjoint representation, and the simplest one of them arises from choosing η to be the orthogonal projector onto $\hat{\mathfrak{h}}$. In that case, by (5.61),

$$\begin{aligned} i\alpha(Z(\eta)) &= -\ \mathrm{trace}\Big[\ \eta\ \mathrm{ad}(H_\alpha)\ \Big] \\ &= -\sum_{\beta\in\Delta}\beta(H_\alpha) - \sum_{\beta\in\hat{\Delta}^-}\beta(H_\alpha) = \sum_{\beta\in\hat{\Delta}^+}\beta(H_\alpha) = \alpha\Big[\ \sum_{\beta\in\hat{\Delta}^+} H_\beta\ \Big] \\ &= 2\alpha(H_{\hat{\delta}}) \end{aligned}$$

for $\alpha\in\Delta$, so $iZ(\eta)$ is the generator in $i\hat{\mathfrak{t}}$ dual to $2\hat{\delta}$, the sum of the positive roots in $\hat{\Delta}$. Moreover, $i\alpha(Z(\eta))$ vanishes for $\alpha\in\Delta'$ (this has been shown above by direct calculation) and is strictly positive for $\alpha\in\hat{\Delta}^+$. (See Sect. 4 of ref. [14] for an explicit proof of the second statement.) Thus $g(\eta)$ is positive definite in this case, and is in fact a *Kähler-Einstein metric* on M. (See Sect. 4 of ref. [14].)

This simple construction can be generalized, as follows. First of all, select a generator $Z_0(\eta)$ in $\hat{\mathfrak{t}}$ lying in the closure of the Weyl chamber $\hat{C}$ in $\hat{\mathfrak{t}}$ that was introduced to represent the given complex structure on M (see (5.54)-(5.56)), and repeat the constructions preceding step 3 with Z_0 replaced by $Z_0(\eta)$. Thus in particular, $H(\eta)$ will be the stability group of $Z_0(\eta)$ under the adjoint representation of G, with Lie algebra $\mathfrak{h}(\eta)$, $\hat{T}(\eta)$ will be its centre, with Lie algebra $\hat{\mathfrak{t}}(\eta)$, and the root system Δ will be split into the disjoint union of the two subsets $\Delta'(\eta)$ and $\hat{\Delta}(\eta)$ given by the analogues of (5.36) and (5.37), respectively, so that the complexified stability algebra $\mathfrak{h}^c(\eta)$ and its orthogonal complement $\mathfrak{m}^c(\eta)$ are given by the analogues of (5.38) and (5.39), respectively. Note that $H(\eta)\supset H$, $\mathfrak{h}(\eta)\supset\mathfrak{h}$, $\hat{T}(\eta)\subset\hat{T}$, $\hat{\mathfrak{t}}(\eta)\subset\hat{\mathfrak{t}}$, $\Delta'(\eta)\supset\Delta'$, $\hat{\Delta}(\eta)\subset\hat{\Delta}$, $\mathfrak{h}^c(\eta)\supset\mathfrak{h}^c$, $\mathfrak{m}^c(\eta)\subset\mathfrak{m}^c$, and that equality will hold if and only if $Z_0(\eta)$ is regular in $\hat{\mathfrak{t}}$, i.e., lies in $\hat{C}$, which means that the interesting cases are those where $Z_0(\eta)$ lies on the boundary of $\hat{C}$. Even in that situation, however, it turns out that the invariant ordering in $\hat{\Delta}$ corresponding to the Weyl chamber $\hat{C}$ in $\hat{\mathfrak{t}}$ induces a unique invariant ordering in $\hat{\Delta}(\eta)$ corresponding to a Weyl chamber $\hat{C}(\eta)$ in $\hat{\mathfrak{t}}(\eta)$, as follows:

$$\hat{\Delta}(\eta)^{\pm} = \hat{\Delta}(\eta)\cap\hat{\Delta}^{\pm}\ ,\quad \hat{C}(\eta)^{cl} = \hat{\mathfrak{t}}(\eta)\cap\hat{C}^{cl}\ . \tag{5.84}$$

(The superfix "cl" stands for "closure".)

Namely, let $\alpha\in\Delta'(\eta)$ and $\beta\in\hat{\Delta}(\eta)^{\pm}$ be such that $\alpha+\beta\in\Delta$; then by the analogue of (5.43), $\alpha+\beta\in\hat{\Delta}(\eta)$, and by the analogue of

(5.49), what must be shown in order to prove invariance of the ordering in $\hat{\Delta}(\eta)$ defined by (5.84) is that $\alpha+\beta\epsilon\hat{\Delta}(\eta)^{\pm}$. For the sake of definiteness, suppose $\beta\epsilon\hat{\Delta}(\eta)^{+}$. Then $\beta\epsilon\hat{\Delta}(\eta)$ implies $\beta(Z_0(\eta))\neq 0$, while $\beta\epsilon\hat{\Delta}^{+}$ implies $\beta(Z_0(\eta))\geq 0$, since $Z_0(\eta)$ lies in the closure of $\hat{C}$, so

$$\alpha(Z_0(\eta)) = 0 \ , \ \beta(Z_0(\eta)) > 0 \implies (\alpha+\beta)(Z_0(\eta)) > 0 \ .$$

Once again, since $Z_0(\eta)$ lies in the closure of $\hat{C}$, it follows that there exists a generator Z' in $\hat{\mathfrak{t}}$ which is close enough to $Z_0(\eta)$ so that $(\alpha+\beta)(Z')>0$, and therefore $\alpha+\beta\epsilon\hat{\Delta}$ cannot belong to $\hat{\Delta}^{-}$ but must belong to $\hat{\Delta}^{+}$, so $\alpha+\beta\epsilon\Delta(\eta)^{+}$, q.e.d..

With these invariant orderings and the corresponding complex structures fixed, the parabolic subalgebra $\hat{\mathfrak{h}}(\eta)$ of $\mathfrak{g}^c$ is given by the analogue of (5.61) and generates the parabolic subgroup $\hat{H}(\eta)$ of G^c. Then choosing η to be the orthogonal projector onto $\hat{\mathfrak{h}}(\eta)$, we can repeat the argument given above to conclude that $iZ(\eta)$ is now the generator in $i\hat{\mathfrak{t}}(\eta)$ dual to $2\hat{\delta}(\eta)$, the sum of the positive roots in $\hat{\Delta}(\eta)$, and that $i\alpha(Z(\eta))$ vanishes for $\alpha\epsilon\Delta'(\eta)$ and is strictly positive for $\alpha\epsilon\hat{\Delta}(\eta)^{+}$. Thus $g(\eta)$ is now positive semidefinite, with

$$\begin{aligned} g(\eta)_{\alpha\bar{\alpha}}\ (z=0,\bar{z}=0) &= 0 \quad \text{for} \quad \alpha\epsilon\hat{\Delta}^{+}\cap\Delta'(\eta) \ . \\ g(\eta)_{\alpha\bar{\alpha}}\ (z=0,\bar{z}=0) &> 0 \quad \text{for} \quad \alpha\epsilon\hat{\Delta}^{+}\cap\hat{\Delta}(\eta) \ . \end{aligned} \tag{5.85}$$

Next, let $Z_0(\eta)$ be one of the vectors in the canonical basis $\hat{B}$ of $\hat{\mathfrak{t}}$ that was introduced to represent the given complex structure on M (see (5.53)). Thus, for $1\leq j\leq\hat{r}$ fixed,

$$Z_{0j} = -iH_{\hat{\lambda}_j} \ , \tag{5.86}$$

$\hat{T}_j$ is the one-dimensional subtorus of $T\subset G$ generated by $-iH_{\hat{\lambda}_j}$,

$\hat{\mathfrak{t}}_j$ is the one-dimensional subspace of $\mathfrak{t}\subset\mathfrak{g}$ generated by $-iH_{\hat{\lambda}_j}$,

$$\begin{aligned} \Delta_j' &= \{\ \alpha\epsilon\Delta \ / \ (\alpha,\hat{\lambda}_j) \equiv \mathrm{Kill}(H_\alpha,H_{\hat{\lambda}_j}) = 0\ \} \ , \\ \hat{\Delta}_j &= \{\ \alpha\epsilon\Delta \ / \ (\alpha,\hat{\lambda}_j) \equiv \mathrm{Kill}(H_\alpha,H_{\hat{\lambda}_j}) \neq 0\ \} \ , \end{aligned} \tag{5.87}$$

which means that Δ_j' is the set of all roots in Δ which are linear combinations of the simple roots in B other than α_j,

$$\begin{aligned}\hat{\Delta}_j^+ &= \hat{\Delta}_j \cap \hat{\Delta}^+ \\ &= \{ \alpha\in\Delta \;/\; (\alpha,\hat{\lambda}_k) \equiv \mathrm{Kill}(H_\alpha,H_{\hat{\lambda}_k}) > 0 \quad \text{for } 1\le k\le \hat{r} \} \;, \\ \hat{\Delta}_j^- &= \hat{\Delta}_j \cap \hat{\Delta}^- \\ &= \{ \alpha\in\Delta \;/\; (\alpha,\hat{\lambda}_k) \equiv \mathrm{Kill}(H_\alpha,H_{\hat{\lambda}_k}) < 0 \quad \text{for } 1\le k\le \hat{r} \} \;,\end{aligned} \qquad (5.88)$$

$$\mathfrak{h}_j^c = \mathfrak{t}^c \oplus \sum_{\alpha\in\Delta_j^!}{}^{\oplus} \mathfrak{g}^{(\alpha)} \;, \qquad \mathfrak{m}_j^c = \sum_{\alpha\in\hat{\Delta}_j}{}^{\oplus} \mathfrak{g}^{(\alpha)} \;,$$

$$\mathfrak{m}_j^c = \mathfrak{m}_j^{(1,0)} \oplus \mathfrak{m}_j^{(0,1)} \;,$$

$$\mathfrak{m}_j^{(1,0)} = \sum_{\alpha\in\hat{\Delta}_j^+}{}^{\oplus} \mathfrak{g}^{(\alpha)} \;, \qquad \mathfrak{m}_j^{(0,1)} = \sum_{\alpha\in\hat{\Delta}_j^-}{}^{\oplus} \mathfrak{g}^{(\alpha)} \;, \qquad (5.89)$$

$$\hat{\mathfrak{h}}_j = \mathfrak{h}_j^c \oplus \mathfrak{m}_j^{(0,1)} \;,$$

η_j is the orthogonal projector onto the $\hat{H}$-invariant parabolic subalgebra $\hat{\mathfrak{h}}_j$ of $\mathfrak{g}^c$, with kernel $\mathfrak{m}_j^{(0,1)}$, and K_j is the function

$$\begin{aligned} K_j(z,\bar{z}) &= \ln \det_{\eta_j}\big(\mathrm{Ad}(g(z))^+\, \mathrm{Ad}(g(z))\big) \;, \\ g(z) &= \exp\Big[\sum_{\alpha\in\hat{\Delta}^+} z^\alpha E_\alpha \Big] \;, \end{aligned} \qquad (5.90)$$

giving rise to the (degenerate) tensor g_j according to

$$(g_j)_{\alpha\bar{\beta}}\,(z,\bar{z}) = \frac{\partial^2 K_j(z,\bar{z})}{\partial z^\alpha\, \partial \bar{z}^\beta} \;, \qquad (5.91)$$

with

$$(g_j)_{\alpha\bar{\beta}}\,(z=0,\bar{z}=0) = i\delta_{\alpha\beta}\alpha(Z_j) \quad \text{for } \alpha,\beta\in\hat{\Delta}^+ \;, \qquad (5.92)$$

where iZ_j is the generator in $i\hat{\mathfrak{t}}$ dual to $2\hat{\delta}_j$, the sum of the positive roots in $\hat{\Delta}_j$; moreover, $i\alpha(Z_j)$ vanishes for $\alpha\in\Delta_j^!$ and is strictly positive for $\alpha\in\hat{\Delta}_j^+$. But this means that $\hat{\delta}_j$ is orthogonal to all the simple roots α_k in $B(\Delta)$ with $k\neq j$, and must therefore, according to (5.52), be some positive multiple of the fundamental weight $\hat{\lambda}_j$ in $\hat{B}(\Lambda)$:

$$Z_j = -ia_j H_{\hat{\lambda}_j} \quad \text{with} \quad a_j > 0 \ . \tag{5.93}$$

Hence if g is an arbitrary G-invariant Kähler metric on M, which according to Theorem 2 can be represented by some generator Z in the Weyl chamber $\hat{C}$ in $\hat{t}$ (see (5.62), (5.63)), then writing

$$Z = -i \sum_{j=1}^{\hat{r}} c_j a_j H_{\hat{\lambda}_j} = \sum_{j=1}^{\hat{r}} c_j Z_j \ , \tag{5.94}$$

with strictly positive coefficients c_j, we see that the function

$$K(z,\bar{z}) = \sum_{j=1}^{\hat{r}} c_j K_j(z,\bar{z}) \tag{5.95}$$

will be a Kähler potential for the G-invariant Kähler metric

$$g = \sum_{j=1}^{\hat{r}} c_j g_j \ . \tag{5.96}$$

In conclusion, I want to briefly describe the solution to the classification problem for homogeneous Kähler manifolds M=G/H. For simplicity, I shall suppose G to be compact and simple. Indeed, it is straightforward to reduce the case where G is semisimple to that where G is simple, and the classification of the non-compact homogeneous Kähler manifolds (with G semisimple) can be reduced to that of the compact homogeneous Kähler manifolds (with G semisimple) and that of the hermitian symmetric spaces of the non-compact type, i.e., the bounded symmetric domains; see ref. [14] for more details.

Step 5: Find a pragmatic classification scheme.

The classification scheme in question is based on the notion of a *painted Dynkin diagram*, which is, by definition, simply an ordinary (connected) Dynkin diagram, with some of its vertices painted black and the remaining ones left white. Obviously, every homogeneous Kähler manifold $M=G/H$ (with G compact and simple) gives rise to such a diagram, as follows.

a) The underlying ordinary Dynkin diagram represents the *simple roots* and is the Dynkin diagram of G, with $r=\mathrm{rank}G=\dim\mathfrak{t}$ vertices.

b) The black vertices, or ***black roots***, are the *simple subgroup roots*, and the subdiagram consisting of the black roots only is the Dynkin diagram of H', the semisimple part of the stability group H, with $r'=\mathrm{rank}H'=\dim\mathfrak{t}'$ vertices.

c) The white vertices, or *white roots*, are the *simple coset roots*; there are $\hat{r}=\dim\hat{\mathfrak{t}}$ of them.

d) The stability group H is obtained from the subdiagram consisting of the black roots only ($\rightarrow$ semisimple part H' of H), with an additional $U(1)$-factor for each white root ($\rightarrow$ centre $\hat{T}$ of H).

Example: The painted Dynkin diagram

represents the homogeneous Kähler manifold

$$E_8 \;/\; SO(8)xSU(3)xU(1)^2 .$$

Conversely, it can be shown that any painted Dynkin diagram does correspond to some homogeneous Kähler manifold, and so the only question that remains is: when are two different painted Dynkin diagrams equivalent in the sense that they determine the same homogeneous Kähler manifold? The answer depends, of course, on what one means by "the same". One natural interpretation is to call two painted Dynkin diagrams equivalent if the corresponding homogeneous Kähler manifolds are diffeomorphic as real manifolds and as coset spaces under G, which means that there should exist a diffeomorphism which commutes with the respective G-actions: this *equivalence problem* has been discussed at length in ref. [14], where a complete list of all painted Dynkin diagrams, up to equivalence in this sense, has been given. The other natural interpretation would be to require that there should even exist a biholomorphic diffeomorphism which commutes with the respective G-actions: this would amount to classifying all (equivalence classes of) complex structures on a given adjoint orbit.

Some results in this direction can be found in the classical paper [27] (Sect. 13.7-13.9) by Borel and Hirzebruch, but for the general case, an intuitively appealing answer (which could, e.g., be read off directly from the corresponding painted Dynkin diagrams) has, to my knowledge, not yet been given.

Note added: After completing this review, I became aware of the paper by Alekseevskii and Perelomov [30] which contains results largely identical with those presented in ref. [14] and in Sect. 5 above (including a discussion of Kähler potentials), and which I recommend for supplementary reading.

References

1. M. Gell-Mann and M. Lévy: The axial vector current in beta decay. Nuov. Cim. 16 (1960) 705-726.

2. S. Weinberg: Nonlinear realizations of chiral symmetry. Phys. Rev. 166 (1968) 1568-1577.

3. B.W. Lee: Chiral dynamics. Gordon & Breach, New York 1972.

4. C. Itzykson and J.B. Zuber: Quantum field theory. McGraw-Hill, New York 1980.

5. M. Forger: Nonlinear sigma models on symmetric spaces, in: Nonlinear partial differential operators and quantization procedures, proceedings, Clausthal, Germany 1981, edited by S.I. Andersson and H.D. Doebner. In: Lecture Notes in Mathematics, Vol. 1037, Springer, Berlin 1983.

6. C. Quigg: Gauge theories of the strong, weak and electromagnetic interactions. Benjamin/Cummings, Reading, Mass. 1983.

7. W. Buchmüller, S.T. Love, R.D. Peccei and T. Yanagida: Quasi-Goldstone fermions. Phys. Lett. 115B (1982) 233-236.
W. Buchmüller, R.D. Peccei and T. Yanagida: Quasi-Nambu-Goldstone fermions. Nucl. Phys. B227 (1983) 503-546.
W. Buchmüller, R.D. Peccei and T. Yanagida: Weak interactions of Quasi-Nambu-Goldstone fermions. Nucl. Phys. B231 (1984) 53-64.
W. Buchmüller, R.D. Peccei and T. Yanagida: The structure of weak interactions for composite quarks and leptons. Nucl. Phys. B244 (1984) 186-206.
W. Buchmüller: Some aspects of supersymmetric composite models of quarks and leptons. Act. Phys. Pol. B16 (1985) 613-626.

8. C.L. Ong: Spontaneously broken supersymmetric systems of the nonlinear fields and gauge fields. Phys. Rev. D27 (1983) 911-916.
C.L. Ong: Gauged, supersymmetric, generalized nonlinear σ models for quarks and leptons. Phys. Rev. D27 (1983) 3044-3047.

9. W. Lerche: On Goldstone fields in supersymmetric theories. Nucl. Phys. B238 (1984) 582-600.

10. G. Girardi, P. Sorba and R. Stora: Comments on the spontaneous symmetry breaking in supersymmetric theories. Phys. Lett. 144B (1984) 212-214.

11. T. Kugo and T. Yanagida: Unification of families based on a coset space $E_7/SU(5)xSU(3)xU(1)$. Phys. Lett. 134B (1984) 313-317.
Y. Achiman, S. Aoyama and J. van Holten: The non-linear supersymmetric σ model on $E_6/SO(10)xU(1)$. Phys. Lett. 141B (1984) 64-68.
Y. Achiman, S. Aoyama and J. van Holten: Gauged supersymmetric σ models and $E_6/SO(10)xU(1)$. Nucl. Phys. B258 (1985) 179-214.
C.L. Ong: Supersymmetric models for quarks and leptons with nonlinearly realized E_8 symmetry. Phys. Rev. D31 (1985) 3271-3279.
S. Irié and Y. Yasui: Supersymmetric non-linear σ model on $E_8/SO(10)xSU(3)xU(1)$. Z. Phys. C29 (1985) 123-126.
W. Buchmüller and O. Napoly: Exceptional coset spaces and the spectrum of quarks and leptons. Phys. Lett. 163B (1985) 161-166.

12. M. Bando, T. Kuramoto, T. Maskawa and S. Uehara: Structure of non-linear realization in supersymmetric theories. Phys. Lett. 138B (1984) 94-98.
M. Bando, T. Kuramoto, T. Maskawa and S. Uehara: Non-linear realization in supersymmetric theories I, II. Progr. Theor. Phys. 72 (1984) 313-349, 1207-1213.

13. K. Itoh, T. Kugo and H. Kunitomo: Supersymmetric nonlinear realization for arbitrary Kählerian coset space G/H. Nucl. Phys. B263 (1986) 295-308.

14. M. Bordemann, M. Forger and H. Römer: Homogeneous Kähler manifolds. Paving the way towards new supersymmetric sigma models. Commun. Math. Phys. 102 (1986) 605-647.

15. W. Buchmüller and W. Lerche: Geometry and anomaly structure of supersymmetric σ models. Ann. Phys. 175 (1987) 159-196.

16. R. Gatto and G. Sartori: Consequences of the complex character of the internal symmetry in supersymmetric theories. Commun. Math. Phys. 109 (1987) 327-352.

17. B. Zumino: Supersymmetry and Kähler manifolds. Phys. Lett. 87B (1979) 203-206.

18. L. Alvarez-Gaumé and D.Z. Freedman: Kähler geometry and the renormalization of supersymmetric σ models. Phys. Rev. D22 (1980) 846-853.
L. Alvarez-Gaumé and D.Z. Freedman: Geometrical Structure and ultraviolet finiteness in the supersymmetric σ model. Commun. Math. Phys. 80 (1981) 443-451.

19. J. Bagger: Supersymmetric sigma models, in: Supersymmetry, proceedings, Bonn, Germany 1984, edited by K. Dietz, R. Flume, G. von Gehlen and V. Rittenberg. Plenum Press, New York 1985.
M. Rocek: Supersymmetry and nonlinear σ models. Physica 15D (1985) 75-82.

20. A. Karlhede, U. Lindström and M. Rocek: Hyperkähler manifolds and nonlinear supermultiplets. Commun. Math. Phys. 108 (1987) 529-534.
N.J. Hitchin, A. Karlhede, U. Lindström and M. Rocek: Hyperkähler metrics and supersymmetry. Commun. Math. Phys. 108 (1987) 535-589.

21. L. Alvarez-Gaumé and P. Nelson: Riemann surfaces and string theories, in: Supersymmetry, supergravity and superstrings, proceedings, Trieste, Italy 1986, edited by B. de Wit, P. Fayet and M. Grisaru. World Scientific, Singapore 1986.
P. Nelson: Lectures on strings and moduli space. Phys. Rep. 149 (1987) 337-375.

22. D. Burns: Some examples of the twistor construction. University of Michigan, Math. Dept. preprint (1985).

23. R. Abraham and J.E. Marsden: Foundations of Mechanics, 2nd edition. Benjamin/Cummings, Reading, Mass. 1978.

24. S. Helgason: Differential geometry, Lie groups, and symmetric spaces. Academic Press, New York 1978.

25. A. Borel: Kählerian coset spaces of semisimple Lie groups. Proc. Nat. Acad. Sci. USA 40 (1954) 1147-1151.

26. H.C. Wang: Closed manifolds with a homogeneous complex structure. Am. J. Math. 76 (1954) 1-32.

27. A. Borel and F. Hirzebruch: Characteristic classes and homogeneous spaces I. Am. J. Math. 80 (1958) 458-538.

28. S. Kobayashi and K. Nomizu: Foundations of differential geometry I, II. Interscience, New York 1963(I)&1969(II).

29. J.E. Humphreys: Linear algebraic groups. Springer, Berlin 1975.

30. D.V. Alekseevskii and A.M. Perelomov: Invariant Kähler-Einstein metrics on compact homogeneous spaces. Funct. Anal. Appl. 20 (1987) 171-183.

NONLINEAR REALIZATIONS, KÄHLER MANIFOLDS AND THE VIRASORO MANIFOLD

Julius Wess
Institut für Theoretische Physik
Universität Karlsruhe

1. Nonlinear Realizations

In the first lecture, I would like to present some technical details that should clarify the notation and demonstrate a physicist's approach to a mathematical problem. A physicist has some specific system in mind - it use to be defined by nature - there he really wants to do some concrete calculations that finally should produce some numbers.

In the next lectures, we shall use the methods to discuss the geometry of the Virasoro-manifold. This part is based on work done with Bruno Zumino at CERN this summer.

We start from a Lie algebra $\underline{G}$ and a subalgebra $\underline{H}$ and we are interested in the manifold G/H where G and H are the respective Lie groups. An element of G will be denoted by g and an element of H by h. The decomposition characterizing the coset G/H is

$$g = mh \qquad g \in G, \quad h \in H \tag{1.1}$$

Let us denote the generators of $\underline{G}$ that are not in $\underline{H}$ by $\underline{M}$, then m should be parametrized as follows:

$$m = e^{\varphi} , \qquad \varphi \in \underline{M} \tag{1.2}$$

The quantity $m^{-1}\,dm$ has a decomposition in $\underline{G}$:

$$m^{-1}\,dm = Q + P , \qquad Q \in \underline{H} , \qquad P \in \underline{M} \tag{1.3}$$

We want to know how Q and P transform under a group motion:

$$g_0 m = m'h' \qquad (1.4)$$

We calculate:

$$\begin{aligned} m'^{-1}\, dm' &= h'm^{-1}\, d\left(mh'^{-1}\right) \\ &= h'\left(m^{-1}\, dm\right)\, h'^{-1} + h'dh'^{-1} \\ &= P' + Q' \end{aligned} \qquad (1.5)$$

We obtain a comparatively simple case if

$$[\, \underline{H}, \underline{M}\,] \in \underline{M} \quad . \qquad (1.6)$$

This, e.g. will be the case for finite dimensional, semi-simple, compact Lie groups. In this case, we find

$$\begin{aligned} P' &= h'\, P\, h'^{-1} \\ Q' &= h'\, Q\, h'^{-1} + h'\, dh'^{-1} \quad . \end{aligned} \qquad (1.7)$$

The quantity P can be interpreted as a "Vielbein" form

$$P = d\lambda^m\, P_m{}^r\, A_r = P^r\, A_r \qquad (1.8)$$

where λ^m are the parameters of the coset space and A_r are the generators in $\underline{M}$. We see that the Vielbein forms P^r transform under a specific representation of H. If we can form a bilinear invariant with this representation, then we can define a metric on the coset space.

$$g_{mn} = P_m{}^r\, P_n{}^s\, \eta_{rs} \, , \qquad (1.9)$$

This, e. g. will be the case if (6) holds and if there exist a Cartan Killing metric.
It follows from the construction that the Killing vectors of g_{mn} include the vector fields that generate the group motions of G.

We are interested in the case that the metric, thus obtained, belongs to a Kähler manifold.

The condition for this to be true will be specified at the end of this lecture.
Let me however first summarize a few properties of a Kähler manifold. A Kähler manifold is a complex manifold whose metric can be obtained from a Kähler potential $K(z,\bar{z})$:

$$g_{i\underline{k}} = \frac{\partial}{\partial z^{i}} \frac{\partial}{\partial \bar{z}^{\underline{k}}} K\left[z,\bar{z}\right] \tag{1.10}$$

With this metric we associate a covariant derivative:

$$\begin{aligned} \nabla_k V_i &= \partial_k V_i + \Gamma_{ki}{}^{l} V_l \,, & \bar{\nabla}_{\underline{k}} V_i &= \bar{\partial}_{\underline{k}} V_i \\ \bar{\nabla}_{\underline{k}} \bar{V}_{\underline{i}} &= \bar{\partial}_{\underline{k}} \bar{V}_{\underline{i}} + \bar{\Gamma}_{\underline{k}\underline{i}}{}^{\underline{l}} \bar{V}_{\underline{l}} \,, & \nabla_k \bar{V}_{\underline{i}} &= \partial_k \bar{V}_{\underline{i}} \end{aligned} \tag{1.11}$$

with the property:

$$\nabla_l \, g_{i\underline{k}} = 0 \qquad\qquad \bar{\nabla}_l \, g_{i\underline{k}} = 0 \tag{1.12}$$

From (12) follows:

$$\begin{aligned} \Gamma_{li}{}^{t} &= - g^{\underline{k}t} \, \partial_l \, g_{i\underline{k}} = \Gamma_{il}{}^{t} \\ \bar{\Gamma}_{\underline{l}\underline{i}}{}^{\underline{k}} &= - g^{\underline{k}t} \, \bar{\partial}_{\underline{l}} \, g_{t\underline{i}} = \bar{\Gamma}_{\underline{i}\underline{l}}{}^{\underline{k}} \end{aligned} \tag{1.13}$$

A curvature can be defined:

$$\begin{aligned} [\nabla_i, \nabla_k] \, V_l &= R_{ikl}{}^{t} \, V_t \\ [\nabla_i, \bar{\nabla}_{\underline{k}}] \, V_l &= R_{i\underline{k}l}{}^{t} \, V_t \\ [\bar{\nabla}_{\underline{i}}, \bar{\nabla}_{\underline{k}}] \, V_l &= R_{\underline{i}\underline{k}l}{}^{t} \, V_t \end{aligned} \tag{1.14}$$

The only nonvanishing component of this curvature is

$$R_{i\underline{k}l\underline{t}} = - g_{t\underline{t}} \, \bar{\partial}_{\underline{k}} \, \Gamma_{il}{}^{t} \tag{1.15}$$

(we raise and lower the indices with the metric).

$$R_{i\underline{k}l\underline{t}} = \partial_i \, \bar{\partial}_{\underline{k}} \, g_{l\underline{t}} - \left(\bar{\partial}_{\underline{k}} \, g_{t\underline{t}}\right)\left(\partial_i \, g_{l\underline{n}}\right) \, g^{\underline{n}t} \tag{1.16}$$

A Ricci tensor can be defined:

$$R_{i\underline{k}} = R_{l\underline{k}i\underline{t}} \, g^{\underline{t}l} = R_{i\underline{k}l\underline{t}} \, g^{\underline{t}l} \tag{1.17}$$

This is correct because the curvature tensor of a Kähler manifold has the following symmetry properties:

$$R_{i\underline{k}l\underline{t}} = R_{l\underline{k}i\underline{t}} = R_{i\underline{t}l\underline{k}} \tag{1.18}$$

A short calculation shows that the Ricci tensor of a Kähler manifold can also be written as:

$$R_{i\underline{k}} = \partial_i \, \bar{\partial}_{\underline{k}} \, \mathrm{tr} \ln g = \partial_i \, \bar{\partial}_{\underline{k}} \ln \det g \tag{1.19}$$

In this form, it is easy to show that the Ricci tensor is proportional to the curvature of a line bundle.
Let us study a line bundle with our Kähler manifold as base space. The transformation properties of such a line bundle are:

$$\Psi(z) = (\det A)^n \, \Psi'(z') \, , \qquad A_i{}^l = \frac{\partial z'^l}{\partial z^i} \tag{1.20}$$

The determinant of the metric tensor transforms correspondingly:

$$\det g(z) = \det A \det \bar{A} \det g'(z') \tag{1.21}$$

The line bundle admits an invariant hermitean metric:

$$\bar{\Psi} \, \Psi \, (\det g)^{-n} = \bar{\Psi}' \, \Psi' \, (\det g')^{-n} \tag{1.22}$$

We define a covariant derivative:

$$\begin{aligned} \nabla_i \Psi &= \partial_i \Psi + n\Gamma_i \Psi \\ \bar{\nabla}_{\underline{k}} \Psi &= \bar{\partial}_{\underline{k}} \Psi \end{aligned} \tag{1.23}$$

and we demand that the covariant derivative be compatible with the metric:

$$\nabla_i \det g = \partial_i \det g + \Gamma_i \det g \equiv 0 \qquad (1.24)$$

This can be solved for Γ_i

$$\Gamma_i = -(\det g)^{-1} \partial_i \det g = -\partial_i \ln \det g \qquad (1.25)$$

We find:

$$\nabla_i \Psi = \partial_i \Psi - n\, \partial_i (\ln \det g)\Psi$$
$$\bar\nabla_{\underline{k}} \Psi = \bar\partial_{\underline{k}} \Psi \qquad (1.26)$$

We obtain the curvature as the commutator of the covariant derivatives:

$$R_{i\underline{k}}\, \Psi = (\nabla_i \bar\nabla_{\underline{k}} - \bar\nabla_{\underline{k}} \nabla_i)\, \Psi = n\, (\bar\partial_{\underline{k}}\, \partial_i \ln \det g)\, \Psi \qquad (1.27)$$

The curvature of the line bundle is n times the Ricci tensor of the base manifold.

We have to know the second derivative of the metric tensor to calculate the curvature. If there is an isometry on the manifold, it is however possible to calculate the curvature along this isometry by using first derivatives only. The relevant concept is the Toeplitz operator. As this procedure facilitates concrete calculations considerably, I would like to discuss this technique in this lecture.

We start from a Lie derivative along a curve that is generated by an isometry, i. e. the vector field defining the Lie derivative is a Killing vector. On a Kähler manifold the equation for a Killing vector is:

$$f^{a}\partial_{a} g_{b\underline{c}} + \bar f^{\underline{a}} \bar\partial_{\underline{a}} g_{b\underline{c}} + \partial_b f^{a} g_{a\underline{c}} + \bar\partial_{\underline{c}} \bar f^{\underline{a}} g_{b\underline{a}} = 0 \qquad (1.28)$$

It is easy to verify that the Lie derivative of a vector field is a vector field again:

$$\pounds_f V_l = \left(f^a \partial_a + \bar{f}^{\underline{a}} \bar{\partial}_{\underline{a}}\right) V_l + \partial_l f^a V_a \tag{1.29}$$

tranforms as a vector field.

Thus we know how to define the commutator of a Lie derivative with a covariant derivative.
The covariant derivative was defined as follows:

$$\nabla_i V_l = \partial_i V_l + \Gamma_{il}{}^k V_k$$
$$\bar{\nabla}_{\underline{k}} V_l = \bar{\partial}_{\underline{k}} V_l \tag{1.30}$$

We compute:

$$\pounds_f \bar{\nabla}_{\underline{k}} V_l = \left(f^a \partial_a + \bar{f}^{\underline{a}} \bar{\partial}_{\underline{a}}\right) \bar{\partial}_{\underline{k}} V_l + \partial_l f^a \bar{\partial}_{\underline{k}} V_a + \bar{\partial}_{\underline{k}} \bar{f}^{\underline{a}} \bar{\partial}_{\underline{a}} V_l \tag{1.31}$$

and

$$\bar{\nabla}_{\underline{k}} \pounds_f V_l = \bar{\partial}_{\underline{k}} [f^a \partial_a V_l + \bar{f}^{\underline{a}} \bar{\partial}_{\underline{a}} V_l + \partial_l f^a V_a] \tag{1.32}$$

It is easy to see that:

$$[\pounds_f, \bar{\nabla}_{\underline{k}}] V_l = 0 \tag{1.33}$$

The same is true for L_f, ∇_i. To verify it, use has been made of equation (28) and (13).

$$[\pounds_f, \nabla_i] V_l = 0 \tag{1.34}$$

We now take the covariant derivative along a vector field g^a

$$\nabla_g = g^a \nabla_a + \bar{g}^{\underline{a}} \bar{\nabla}_{\underline{a}} \tag{1.35}$$

The curvature along two vector fields can be seen to be

$$R_{fg} = [\nabla_f, \nabla_g] - \nabla_{\{f,g\}} \tag{1.36}$$

with

$$\{f,g\} = \left(f^a \partial_a g^b - g^a \partial_a f^b\right)\partial_b + \left(\bar{f}^{\underline{a}} \bar{\partial}_{\underline{a}} \bar{g}^{\underline{b}} - \bar{g}^{\underline{a}} \bar{\partial}_{\underline{a}} \bar{f}^{\underline{b}}\right)\bar{\partial}_{\underline{b}} \; . \tag{1.37}$$

We calculate the commutator between the Lie derivative and the covariant derivative, both along a Killing vector field and we find:

$$[\ell_f, \nabla_g] = \nabla_{\{f,g\}} \tag{1.38}$$

This follows from (33) and (34) and the definition of the Lie derivatives.

The Toeplitz operator is defined as follows:

$$T_f = \ell_f - \nabla_f \tag{1.39}$$

We calculate the commutator

$$[T_f, T_g] = R_{fg} + T_{\{f,g\}} \; . \tag{1.40}$$

The curvature can be expressed in terms of the Toeplitz operator. It is important to note, that, due to the definition, the Toeplitz operator does not involve any derivatives.

When we derive a metric on a group manifold with the techniques discussed before, we might be able to impose a complex structure. To find out if it also allows a Kähler structure, we should be able to find the Kähler potential. For this potential to exist, the metric has to satisfy integrability conditions:

$$\partial_b g_{a\underline{c}} = \partial_a g_{b\underline{c}} \qquad\qquad \bar{\partial}_{\underline{b}} g_{a\underline{c}} = \bar{\partial}_{\underline{c}} g_{a\underline{b}} \tag{1.41}$$

This, of course, is equivalent to demand that the Kähler-form should be closed. In this lecture, however, I would like to discuss this condition without using forms. We assume that there are three vector fields f_A, f_B and f_C. From the integrability conditions then follows:

$$\oint_{\substack{ABC\\ \text{cyclic}}} \left(f_A{}^b\partial_b + \bar{f}_A{}^{\underline{b}}\bar{\partial}_{\underline{b}}\right) g_{a\underline{c}} \left(f_B{}^a\bar{f}_C{}^{\underline{c}} - f_C{}^a\bar{f}_B{}^{\underline{c}}\right) = 0 \tag{1.42}$$

If the vector fields are Killing vector fields, satisfying equation (28), we derive from (42)

$$\oint_{\substack{ABC\\ \text{cyclic}}} \left(\partial_a f_A{}^b g_{b\underline{c}} + \bar{\partial}_{\underline{c}}\bar{f}_A{}^{\underline{b}} g_{a\underline{b}}\right) \left(f_B{}^a\bar{f}_C{}^c - f_C{}^a\bar{f}_B{}^{\underline{c}}\right) = 0 \tag{1.43}$$

and consequently:

$$\oint_{\substack{ABC\\ \text{cyclic}}} \left[\{f_A,f_B\}^b\,\bar{f}_C{}^{\underline{c}} - \{\bar{f}_A,\bar{f}_B\}^{\underline{c}}\,f_C{}^b \right] g_{b\underline{c}} = 0 \tag{1.44}$$

This is the condition the metric has to satisfy.
The manifold, according to (4) is parametrized as follows:

$$m = e^{x^l A_l} \qquad A_l \in \underline{M} \tag{1.45}$$

The group element g_0 has parameters λ^l

$$g_0 = e^{\lambda^l T_l} \qquad T_l \in \underline{G}$$
$$[T_l,T_k] = f_{lk}{}^r T_r \tag{1.46}$$

and it moves the manifold as follows:

$$e^{\lambda^l T_l}\, e^{x^l A_l} = e^{x'^l A_l}\, e^{b^l V_l} \qquad V_l \in \underline{H} \tag{1.47}$$

For each parameter λ^l, a curve is generated:

$$\lambda^r f_r{}^l = x'^l - x^l \tag{1.48}$$

The Lie derivative along this curve is

$$\ell_r = f_r{}^l \frac{\partial}{\partial x^l} \tag{1.49}$$

Near the origin, the vector fields can be easily calculated and it follows that the Lie derivatives satisfy the relation:

$$[L_r, L_s] = L_{\{f_r f_s\}} = - f_{rs}{}^t \, L_t \tag{1.50}$$

This can be combined with (44) to derive the Kähler condition for the metric at the origin.

2. The Virasoro Manifold

We proceed according to the previous lecture and compute the Ricci tensor of the Virasoro maniold.
The Virasoro algebra, Diff S_1, is defined as follows:

$$[L_n, L_m] = (n - m)\, L_{n+m} \qquad n,m \in Z \tag{2.1}$$

With this algebra the following conjugation relation is consistent:

$$L_m^+ = L_{-m} \tag{2.2}$$

We define as Virasoro manifold Diff S_1/L_0. This manifold allows a complex structure.

$$u = e^{\{\alpha^m L_m - \bar{\alpha}^m L_{-m}\}} \, e^{\rho L_0} \quad , \qquad m > 0 \tag{2.3}$$

Here we assume $\bar{\rho} = - \rho$.
This, however, is not a very suitable parametrization for the Kähler structure.
The following parametrization is more convenient:

$$u = e^{z^m L_m} \; e^{\mu^m L_{-m}} \; e^{\chi L_0} \qquad m > 0 \tag{2.4}$$

We have to express μ in terms of z, $\bar{z}$ to match (2,3):

$$\mu^r = -\bar{z}^r + (r + 2n)\, z^n \bar{z}^{r+n} + \ldots \qquad r > 0 \tag{2.5}$$

Summation over n>0 is understood. The dots denote, and do so in the future, higher powers in z,$\bar{z}$.
To find a "Vielbein" on the Virasoro manifold, at least near the origin z = $\bar{z}$ = 0, we use equation (1,3)

$$m^{-1}\, dm = e^{-\mu^n L_{-n}}\, e^{-z^n L_n}\, d(e^{z^n L_n}\, e^{\mu^n L_{-n}})$$

$$= e^{-\mu^n L_{-n}} \{e^{-z^n L_n}\, de^{z^n L_n}\}\, e^{\mu^n L_{-n}} + e^{-\mu^n L_{-n}} de^{\mu^n L_{-n}}$$

$$= P^+ + P^- + P^0 \quad . \tag{2.6}$$

P^+, P^-, P^0 lie in the subalgebras with positive, negative or zero on the generators L respectively. We note that P^+ can only arise from the first terms in (2,6) and, therefore, involves dz only (not d$\bar{z}$). The same is true for P^0.
We use Hausdorff's formula

$$e^A e^B = e^{A + B + \frac{1}{2}[AB] + \ldots}$$

to this order to compute P^+ and P^0. This is a valid approximation near the origin.

$$P^+ = dz^r \{L_r + (2r - s)\left(\tfrac{1}{2} z^{s-r} - \bar{z}^{r-s}\right) L_s + \ldots \}$$

$$P^0 = -2r\bar{z}^r dz^r + \ldots \tag{2.7}$$

For our algebra (2,1) the condition (1,6) holds (H = L_0). Equation (1,7) is therefore applicable. The Vielbein-form can be deduced from (2,7):

$$P_+^s = dz^m \{\delta_m^s + (2m - s)\left(\tfrac{1}{2} z^{s-m} - \bar{z}^{m-s}\right) + \ldots \} \tag{2.8}$$

We now know from (1,7) and the definition of the algebra (2,1) that each P_+^s transforms with a phase and that

$$\sum_{s>0} f(s) \left(P_+^s\right)^+ P_+^s \tag{2.9}$$

is an invariant for any function f(s).
To find a Kähler structure, we have to impose the condition (1,44) in addition. At the origin ($z = \bar{z} = 0$) this condition takes a simple form and has the solution:

$$f(s) = as^3 + bs \tag{2.10}$$

were a and b are arbitrary constants.
This yields the metric:

$$\begin{aligned} g_{m\underline{n}} &= (am^3 + bm)\,\delta_{mn} \\ &\quad + [(an^3 + bn)\left(m - \tfrac{n}{2}\right) - (am^3 + bm)(2n - m)]\, z^{n-m} \\ &\quad + [(am^3 + bn)\left(n - \tfrac{m}{2}\right) - (an^3 + bn)(2m - n)]\, \bar{z}^{m-n} \\ &\quad + \ldots \end{aligned} \tag{2.11}$$

The corresponding Kähler potential is:

$$\begin{aligned} K &= (am^3 + bm)\, z^m \bar{z}^m \\ &\quad + \tfrac{1}{2}\, [(as^3 + bs)\left(r - \tfrac{s}{2}\right) - (ar^3 + br)(2s - r)][z^s \bar{z}^r \bar{z}^{s-r} + \bar{z}^s z^r z^{s-r}] \\ &\quad + \ldots \end{aligned} \tag{2.12}$$

To compute the Ricci tensor for the Virasoro manifold, we follow the procedure outlined in the last lecture that uses the Toeplitz operator.

This has the advantage that we have to know the metric only to the first order in z, $\bar{z}$ to compute the curvature tensor at the origin. First we have to derive the Killing operators of the metric (2,11) that are associated with the group motion. Equation (1,4) tells us how to compute this motion.
The answer is:

$$z'^r = z^r + \alpha^r + \frac{1}{2}(2m - r)\, \alpha^m z^{r-m} + (2m + r)\, \bar{\alpha}^m z^{r+m} - r\rho z^r + \ldots \qquad (2.13)$$

The parameters α and ρ have been introduced in (2,3).
The derivative part of the Lie derivaties along these curves are:

$$\mathcal{L}_\alpha = \{\alpha^r + \frac{1}{2}(2m - r)\, \alpha^m z^{r-m} + (2m + r)\, \bar{\alpha}^m z^{r+m}\} \frac{\partial}{\partial z^r} + \{\bar{\alpha}^r + \frac{1}{2}(2m - r)\, \bar{\alpha}^m \bar{z}^{r-m} + (2m + r)\, \alpha^m \bar{z}^{r+m}\} \frac{\partial}{\partial \bar{z}^r} \qquad (2.14)$$

We also now from equation (1,13) how to compute the covariant derivatives compatible with the metric (2,11):

$$\Gamma_{il}{}^t = \Gamma_{li}{}^t = - \frac{\delta^t_{l+i}}{at^3+bt} \{(at^3 + bt)(i - \frac{t}{2}) - (ai^3 + bi)(2t - i)\} \qquad (2.15)$$

The Toeplitz operator is now easy to compute.

$$T_{\alpha l}{}^k = - \delta_\alpha z^i\, \Gamma_{il}{}^k + \partial_l\, \delta_\alpha\, z^k$$

$$T_{\rho l}{}^k = - \delta_\rho\, z^i\, \Gamma_{il}{}^k + \partial_l\, \delta_\rho\, z^k \qquad (2.16)$$

The operators that are generated by α^m, $\bar{\alpha}^m$, ρ we denote by T_m, T_{-m} and T_0 respectively. Note that $\bar{\alpha}^m$ enters with a negative sign in (2.3).

$$T_{m,l}{}^{k} = -\delta_{m+l}{}^{k}\frac{(al^3 + bl)(2k - 1)}{ak^3 + bk}$$

$$T_{-m,l}{}^{k} = \delta_{l}{}^{m+k}(2m + k) \tag{2.17}$$

$$T_{0,l}{}^{k} = -l\delta_{l}{}^{k} \qquad\qquad l,\ m,\ k > 0.$$

The commutator of two Toeplitz operators is easy to calculate:

$$T_{m,l}{}^{k}\, T_{-n,k}{}^{t} - T_{-n,l}{}^{k}\, T_{m,k}{}^{t} =$$

$$-\delta^{t+n}_{l+m}\left[\frac{al^3 + bl}{a(l+m)^3 + b(l+m)}(2m + l)(l + m + n)\right.$$

$$\left. - \theta(l - n)\theta(t - m)\frac{a(l-n)^3 + b(l-n)}{a(l+m-n)^3 + b(l+m-n)}(n + l)(l + 2m - n)\right]$$

(2.18)

The curvature at the origin is then given by:

$$R_{m,-n,l}{}^{t} = [T_m, T_{-n}]_l{}^{t} - (m + n)\,T_{m-n,l}{}^{t} \tag{2.19}$$

For the computation of the Ricci tensor, we are only interested in the diagonal elements $R_{m,-n(l)}{}^{(l)}$.
They are only different from zero for n = m. We obtain from (2,19), for b = 0

$$R_{m,-m(l)}{}^{(l)} = -\left[(l + 2m)^2\frac{l^3}{(l+m)^3} - 2ml - \theta(l - m)(l + m)^2\frac{(l - m)^3}{l^3}\right]$$

(2.20)

Note that a has dropped out of this formula. The asymptotic behavior of (2,20) is like $1/l^2$, the series is absolute convergent. This would not have been the case had we put a = 0 and kept b ≠ 0.
If we sum over l, as we have to do for the Ricci tensor, we can rearrange the summation .

We find:

$$\sum_{l} R_{m,-m\; l}{}^{l} = \lim_{N \to \infty} \left[- \sum_{l=1}^{N} \left[(l + 2m)^2 \frac{l^3}{(l+m)^3} - 2ml \right] \right.$$

$$\left. + \sum_{l=m+1}^{N} (l + m)^2 \frac{(l-m)^3}{l^3} \right] \tag{2.21}$$

In the second sum, we replace l by l - m and obtain

$$\sum_{l=1}^{N} R_{m,-m\; l}{}^{l} = + \frac{1}{2}\, 2mN\, (N + 1) - \sum_{l=N-m+1}^{N} \frac{l^3(l + 2m)^2}{(l + m)^3} \tag{2.22}$$

The remaining sum can be changed to a sum from 1 to m by introducing the summation index $p = l + m - N$, and therefore, will be a finite sum for fixed m:

$$\sum_{l=1}^{N} R_{m,-m\; l}{}^{l} = + mN(N + 1) - \sum_{p=1}^{m} \frac{(N + m + p)^2 (N - m + p)^3}{(N + p)^3} \tag{2.23}$$

The terms that behave like $1/(N+p)$ will tend to zero for $N \to \infty$ (p has a finite range). The finite sums can be calculated and the result is

$$R_{m,-n} = \delta_{m,n} \sum_{l=1}^{\infty} R_{m,-m\; l}{}^{l} = \left[+ \frac{26}{12} m^3 - \frac{1}{6} m \right] \delta_{m,n} \tag{2.24}$$

Would we have kept $a \neq 0$, $b \neq 0$, the result would have been the same. The terms depending on b would have gone to zero in the limit, $N \to \infty$.

The number 26/12 is what we wanted to compute. In a theory, where Diff S_1 is a symmetry group we can consider the vacuum state as a line bundle over the manifold Diff S_1/L_0; L_0 is supposed to be the stability group of the vacuum. To be physically meaningful, the vacuum should have vanishing curvature, the argument is similar as for Barry's phase in quantum mechanics. We conclude that $26/12\, m^3$ of the Ricci tensor has to be compensated by a representation of Diff S_1 in the matter sector. It is well known that this is the case for a bosonic string theory in 26 dimensions.

I would like to thank R. Garreis for his help preparing the manuscript.

Literature

1. S. Coleman, J. Wess, B. Zumino, Phys. Rev., 177, (1969), 2239
2. J. Wess, in Springer Tracts in Modern Physics, 50, (1969)
3. B. Zumino, LBL-23056 (1987) Lawrence Berkeley Laboratory
4. M. J. Bowick and S. G. Rajeev, Phys. Rev. Lett., 58, (1987) 535 and CTP # 1450, Massachussetts Institute of Technology - submittet to Nuclear Physics
5. D. Freed, in Infinite Dimensional Groups with Applications, V. Kac (Springer Verlag Belin, 1985)
6. K. Pilch and N. P. Warner, CTP 1457, Massachussetts Institut of Technology - submitted to Journal of Classical and Quantum Gravity

IN SEARCH OF THE GRADED MANIFOLD OF MAPS BETWEEN GRADED MANIFOLDS

M. Batchelor*
Department of Pure Mathematics and Mathematical Statistics,
University of Cambridge, 16 Mill Lane, Cambridge CB2 1SB

The contribution of the theory of finite dimensional manifolds to theoretical physics is substantial. The lack of an equally well-developed theory of infinite-dimensional manifolds is in part responsible for the divergence of pure mathematics and mathematical physics. The pure mathematicians have not provided a sufficiently convenient framework, and the physicists have had to go it alone.

In these lectures I would like to propose a mathematical framework for discussing the differential topology of infinite dimensional manifolds which are the spaces of smooth maps between smooth finite dimensional manifolds. My plan is to exploit a technique, that of dual coalgebras, which is almost too obvious in the case of ordinary manifolds, has been of demonstrated value in the case of gradual manifolds, and can be generalized to the case of the manifold of smooth maps between smooth manifolds.

The idea behind the construction is as follows. Traditionally there are two approaches to the study of geometry - the direct approach doing analysis on a smooth manifold X and its tangent space TX, or the algebraic approach, studying the ideal structure of $C^\infty(X)$.

The point is that X, TX themselves have algebraic structures which seem perhaps too obvious to state. Both X and TX have a product rule:

For x in X, writing $f(x) = \langle x,f \rangle$, f,g in $(C^\infty(X)$,

$$\langle x, fg \rangle = \langle x\, f \rangle \langle x,g \rangle$$

For a tangent vector ξ_x in $T_x X$, writing $\langle \xi_x, f \rangle = \xi_x f$

$$\langle \xi_x\, fg \rangle = \langle \xi_x, f \rangle \langle x,g \rangle + \langle x,f \rangle \langle \xi_x, g \rangle .$$

Another way of saying this is to say that X, TX live in a subspace $C^\infty(X)^\circ$ of the full linear dual $C^\infty(X)'$ of $C^\infty(X)$, which has a coproduct

$$\Delta : C^\infty(X)^\circ \to C^\infty(X)^\circ \otimes C^\infty(X)^\circ$$

$$\Delta\eta = \sum_i \eta_{(1)i} \otimes \eta_{(2)i} ,$$

(in our examples $\Delta x = x \otimes x$, $\Delta \xi_x = \xi_x \otimes x + x \otimes \xi_x$), such that

$$\langle \eta, fg \rangle = \sum_i \langle \eta_{(1)i}, f \rangle \langle \eta_{(2)i}, g \rangle .$$

This space $C^\infty(X)^\circ$ and its generalizations are the subject of these talks. I have argued that $C^\infty(X)^\circ$ contains X and TX. I will show that it contains the dual bundles to the jet bundles. For ordinary manifolds X and for graded manifolds (X,A), this coalgebra can be derived from the algebra of functions.

* Supported by SERC grant B/AF/683.

For the manifold of smooth maps between manifolds, I will postulate a candidate for the dual coalgebra, and construct functions for the manifold of maps as functions on the dual coalgebra. The organization of the paper is as follows:

I. *The dual coalgebra and ordinary manifolds.*

Definition of the dual coalgebra, structure theorems for $C^\infty(X)^\circ$. The bundles $\bar{T}^k$. Interpretation of Lagrangian densities, polynomial differential operators and geometric prequantization.

II. *Applications of dual coalgebras to graded manifolds.*

Structure theorem and interpretation of $A(X)^\circ$. Graded Lie groups. Building supermanifolds from graded manifolds.

III. *The dual coalgebra and the manifold of smooth maps.*

Definition of $P(B,A)$, and structure theorem for $P(B,A)$. Geometric interpretation for $B = C^\infty(Y), A = C^\infty(X)$. Examples: strings, spinning strings and superstrings.

IV. *Application.*

Berezin integration.

V. *Application.*

The automorphism group of the graded Riemann sphere.

VI. *Application.* The functions on the space of smooth maps. The action integrals for strings are used as examples.

I am indebted to many people: to Peter Bryant who brought measuring coalgebras to my attention; to I.M. Singer who invited me to give a series of seminars at MIT on this subject; to Anna Achucarro for reading the manuscript and advising on the applications; to Sarah Shea-Simonds for typing a difficult manuscript; and last and most of all, to the organiers of this workshop for the opportunity to talk, and for their leniency in enforcing deadlines.

I. The Dual Coalgebra and Ordinary Manifolds

The dual coalgebra is defined for any algebra A over $\mathbb{R}$ (but it could be equally over $\mathbb{C}$ or any other field) in a purely algebraic manner.

I.1 Definition

For any algebra A over $\mathbb{R}$, define A° to be the subspace of the full linear dual A' given by

$$A^\circ = \{\alpha \in A' : \ker \alpha \supset \mathrm{I},$$

$$\mathrm{I} \text{ is an ideal}$$

$$A/\mathrm{I} \text{ is finite dimensional}\}.$$

This is the set of 'almost homomorphisms': clearly α is in A° if and only if α can be written $\alpha = \bar{\alpha}\nu$

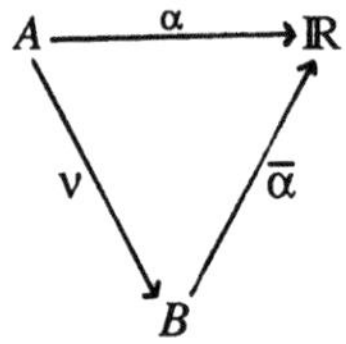

where B is a finite dimensional algebra, ν is an algebra homomorphism, and $\bar{\alpha}$ is linear.

The elementary properties of A° are summarized in the following proposition.

I.2 Proposition

A,B are algebras.

1. A° is a linear subspace of A'
2. $(A \otimes B)^\circ = A^\circ \otimes B^\circ$
3. If $f: A \to B$ is an algebra homomorphism f induces a linear map $f^\circ: B^\circ \to A^\circ$
4. Multiplication $\mu: A \otimes A \to A$ induces a comultiplication

$$\Delta: A^\circ \to A^\circ \otimes A^\circ$$

satisfying

$$<\alpha, ab> = <\Delta\alpha, a \otimes b>.$$

This gives A° the structure of a coalgebra. Moreover the map f of 3 is a coalgebra homomorphism.

5. If A is (graded) commutative, A° is (graded) co-commutative.
6. Every element α in A° is an element of a finite dimensional subcoalgebra.
7. $A^{\circ\prime}$ is an algebra, and there is an algebra homomorphism $A \to A^{\circ\prime}$.

Proof.

1. If α in A° vanishes on I, so does $r\alpha$ for r in $\mathbb{R}$. If α, β are in A°, with α vanishing on I, β vanishing on J, then $\alpha + \beta$ vanishes on $\mathrm{I} \cap \mathrm{J}$. If I, J are cofinite, then I, J is cofinite, since the map

$$A/\mathrm{I} \cap J \to A/\mathrm{I} \oplus A/J$$

is injective, and A/I, A/J are both finite dimensional.

2. Check $A^{\circ} \otimes B^{\circ} \subset (A \otimes B)^{\circ}$. Let $\alpha \in A^{\circ}$, $\beta \in B^{\circ}$, and suppose $\alpha(\mathrm{I}) = 0$, $\beta(J) = 0$, for I, J cofinite ideals in A,B respectively. Then $\mathrm{I} \otimes B + A \otimes J$ is an ideal in $A \otimes B$, the kernel of the map $A \otimes B \to (A/\mathrm{I}) \otimes (B/J)$, and hence $\mathrm{I} \otimes B + A \otimes J$ is cofinite. This shows that $\alpha \otimes \beta$ is in $(A \otimes B)^{\circ}$, and since $A^{\circ} \otimes B^{\circ}$ is spanned by elements $\alpha \otimes \beta \subset (A \otimes B)^{\circ}$.

$$A^{\circ} \otimes B^{\circ} \subset (A \otimes B)^{\circ}.$$

Check $(A \otimes B)^{\circ} \subset A^{\circ} \otimes B^{\circ}$. Let μ be in $(A \otimes B)^{\circ}$, and let μ vanish on the cofinite ideal $H \subset A \otimes B$. Observe that the ideals $A \cap H$, $B \cap H$ are cofinite, and that the map $(A/(A \cap H)) \otimes (B/(B \cap H)) \to A \otimes B/H$ is surjective. Writing $\mu = \bar{\mu} r$,

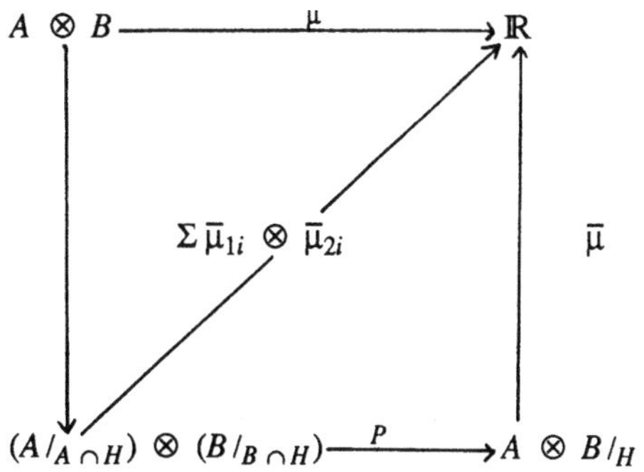

where r is a homomorphism, $\bar{\mu} P$ can be expressed as a sum $\bar{\mu} P = \Sigma \bar{\mu}_{1i} \otimes \bar{\mu}_{2i}$ with $\bar{\mu}_{1i}$ in $(A/(H \cap A))'$, $\bar{\mu}_{2i}$ in $(B/(H \cap B))'$. Now observe that $r_1 : A \to A/(H \cap A)$ is a homomorphism onto a finite dimensional algebra, and hence $\mu_{1i} = \bar{\mu}_{1i} r_1$ is in A°. Similarly construct μ_{2i} in B°, and verify that $\mu = \Sigma \mu_{1i} \otimes \mu_{2i}$ in $A^{\circ} \otimes B^{\circ}$.

3. Define $f^{\circ}\beta = \beta f$ for β in B°. If β vanishes on I, $f^{\circ}\beta$ vanishes on $f^{-1}(\mathrm{I})$. If B/I is finite dimensional, $A/_{f^{-1}(\mathrm{I})}$ is too and $f^{\circ}\beta$ is in A°.

4. Let α be in A°, vanishing on I. As in 2, $A \otimes \mathrm{I} + \mathrm{I} \otimes A$ is a cofinite ideal in $A \otimes A$. Moreover, for $\Sigma f_i \otimes g_i$ in $A \otimes \mathrm{I} + \mathrm{I} \otimes A$, $\Sigma f_i g_i$ is in I. Therefore $<\alpha, \Sigma f_i g_i> = 0 = <\mu' \alpha, \Sigma f_i \otimes g_i>$, where $\mu' : A' \to (A \otimes A)'$. This shows that μ' restricts to a map $\mu^{\circ}: A^{\circ} \to (A \otimes A)^{\circ}$. Using 2 above, define $\Delta = \mu^{\circ}: A^{\circ} \to (A \otimes A)^{\circ} = A^{\circ} \otimes A^{\circ}$.

For A° to be a coalgebra, Δ must be coassociative; i.e.

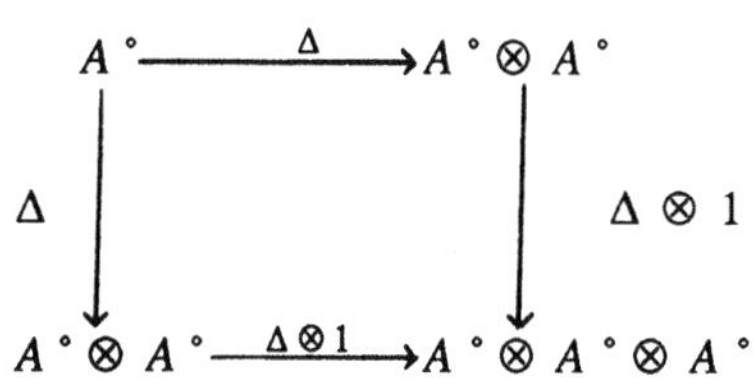

and the counit $\quad \varepsilon : A^{\circ} \to \mathbb{R}, \quad \varepsilon(\alpha) = \langle \alpha, 1 \rangle$

must satisfy

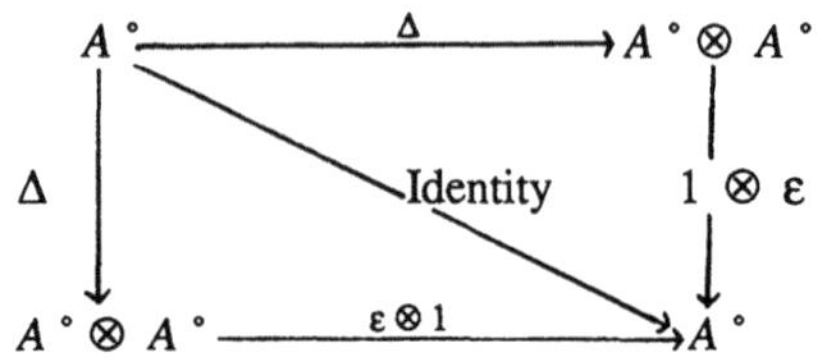

These properties follow from the properties of associativity and units in A.

Similarly, it is not hard to verify that f° is a coalgebra map, i.e. the diagram

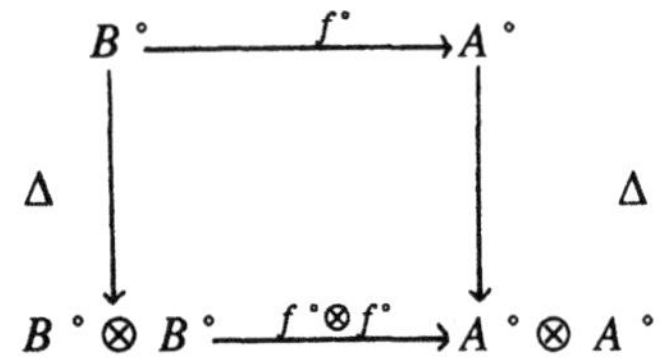

commutes.

5. As with the verification in 4, these are straightforward.

6. This fact holds for any coalgebra C, not just dual coalgebras. Since it is the principal feature which distinguishes coalgebra theory from algebra theory, and the feature which makes the technique appropriate for analyzing badly infinite algebras, the proof is sketched in its generality. See Sweedler, p.46 [7].

Let C be any coalgebra, and C' its full linear dual. Multiplication in C' is defined as in 7. I can consider C as a module for C' under the action

$$\gamma.c = \Sigma \langle \gamma, c_{1i} \rangle \, c_{2i}$$

for γ in C', c in C and $\Delta c = \Sigma c_{1i} \otimes c_{2i}$.

Now suppose c is in C, and construct a finite dimensional subcoalgebra containing c as follows. Let N be the submodule generated by c as a C' module. By the definition of the action of C', it is evident that N is spanned by $\{ c_{2i} \}$, and hence N is finite dimensional.

Let J be the kernel of the algebra homomorphism

$$C' \to \mathrm{End}(N).$$

Thus J is cofinite, and its dual space

$$J^{\perp} = \{ d \in C : \langle d, J \rangle = 0 \}$$

is finite dimensional, and is a subcoalgebra of C.

Finally, observe that c is in $J^{\perp}$:

$$\langle \gamma, c \rangle = \Sigma \langle \gamma, c_{1i} \rangle \, \varepsilon(c_{2i})$$

$$= \varepsilon\ (\Sigma <\gamma, c_{1i}> c_{2i})$$

$$= \varepsilon\ (\gamma . c)\,.$$

If γ is in J, by definition $\gamma . c = 0$, and hence $\varepsilon(\gamma . c) = 0$. Thus c is in $J^{\perp}$.

7. Multiplication in $A^{\circ\prime}$ is given by

$$<pq, \alpha> \;=\; <p \otimes q, \Delta\alpha>\,.$$

Further results about coalgebras in general are given in Sweedler's book *Hopf Algebras*.

Of more interest than general behavioural properties here, is the matter of what $A^{\bullet}$ looks like in the case $A = C^{\infty}(X)$, X is a nice (= paracompact smooth Hausdorff, without boundary) manifold. It is worth introducing some notation.

I.3 Notation

It is not always possible to stick to a consistent convention; thus for a in A, α in A° I will write

$$<a, \alpha> \;=\; <\alpha, a>$$

or in the graded case

$$<a, \alpha> \;=\; (-1)^{|a||\alpha|} <\alpha, a>\,.$$

I will adopt the sigma notation of Sweedler (p.10, [7]), and write

$$\Delta\alpha \;=\; \sum_{(\alpha)} \alpha_{(1)} \otimes \alpha_{(2)}$$

suppressing the index i.

I.4 Notation

Let X be a nice manifold. Define

$$\mathrm{I}_x \;=\; \{f \in C^{\infty}(X) : f(x) = 0\}$$

$$\bar{T}_x^k \;=\; \{\alpha \in C^{\infty}(X)' : \alpha(\mathrm{I}_x^{k+1}) = 0\}$$

Let I be an ideal in $C^{\infty}(X)$. Then

$$\mathrm{I}^{\perp} \;=\; \{\alpha \in C^{\infty}(X)^{\circ} : \alpha(\mathrm{I}) = 0\}\,.$$

I.5 Proposition

1. $\bar{T}_x^k \subset C^{\infty}(X)^{\circ}$,

$\bar{T}_x^k = \left[C^{\infty}(X)/\mathrm{I}_x^{k+1}\right]' = \left(\mathrm{I}_x^{k+1}\right)^{\perp}$, and $\bar{T}_x^k$ is a subcoalgebra, that is,

$\Delta\bar{T}_x^k \subset \bar{T}_x^k \otimes \bar{T}_x^k$

2. $\bar{T}_x^0 = \mathbb{R}x$

3. $\bar{T}_x^1 = T_xX \oplus \mathbb{R}x$

4. $\bar{T}_x^k \subset \bar{T}_x^{k+1}$.

5. $\left(\prod_{j=1}^{Q} \left(\mathrm{I}_{x_j}^{k_j+1}\right)\right)^{\perp} = \bigoplus_{j=1}^{Q} \bar{T}_{x_j}^{k_j}$ $\quad x_j$ all distinct.

Proof.

1. It is sufficient to check that I_x^{k+1} is cofinite in $C^\infty(X)$. Notice that functions f,g represent the same class in $C^\infty(X)/\mathrm{I}_x^{k+1}$ if and only if the Taylor's series expansion of f at x agrees with the Taylor series expansion at g at x up to order $k+1$. Thus $\dim\,(C^\infty(X)/\mathrm{I}_x^{k+1})$ is given by the number of polynomials in $\dim X$ variables of order $\leq k$. The rest of 1 is now immediate.

2. By the definition of I_x, $\mathbb{R}x \subset (\mathrm{I}_x)^\perp = \bar{T}_x^0$. The result follows since $\bar{T}_x^0 = \left[C^\infty(X)/\mathrm{I}_x\right]'$, and since $C^\infty(X)/\mathrm{I}_x$ is one dimensional.

3. Since $\mathrm{I}_x \subset \mathrm{I}_x^2$, $\mathbb{R}x \subset (\mathrm{I}_x^2)^\perp = \bar{T}_x^1$. Check directly (using the product rule) that $T_xX \subset \bar{T}_x^1$. Calculating the dimension of $C'(X)/\mathrm{I}_x^2$ as in 1 above demonstrates that $T_xX \oplus \mathbb{R}x$ is all of $T_x^1 = \left[C^\infty(X)/\mathrm{I}_x^2\right]'$.

4. This is an immediate consequence of the inclusion $\mathrm{I}_x^{k+2} \subset \mathrm{I}_x^{k+1}$.

5. A generalization of the Chinese Remainder Theorem states that the algebra homomorphism

$$r : C^\infty(X) / \bigcap_{j=1}^{Q} \mathrm{I}_{x_j}^{k_j+1} \rightarrow \bigoplus_{j=1}^{Q} C^\infty(X) / \mathrm{I}_{x_j}^{k_j+1}$$

is surjective [4] if for any $j \neq j'$, $\mathrm{I}_{x_j}^{k_j+1} + \mathrm{I}_{x_{j'}}^{k_{j'}+1} = C^\infty(X)$. (See, for example, *Algebra*, P.M. Cohn, p.110.) This can be demonstrated by choosing a function ψ in $C^\infty(X)$ which is 1 in a neighbourhood of x_j and 0 in a neighbourhood of $x_{j'}$. Observe that ψ is in $\mathrm{I}_{x_{j'}}$, $1-\psi$ is in I_{x_j}, and by expanding $1 = (\psi + (1-\psi))^{k_j + k_{j'} + 2}$ by the binomial expansion yields the result that 1 is in $\mathrm{I}_{x_j}^{k_j+1} + \mathrm{I}_{x_{j'}}^{k_{j'}+1}$. Thus the condition is satisfied and r is an isomorphism.

The result now follows from Part 1.

So far X and TX have been identified as subsets of $C^\infty(X)^\circ$. A full description of $C^\infty(X)^\circ$ can now be given.

I.6

THE STRUCTURE OF $C^\infty(X)^\circ$ - PROBABLY THE MOST IMPORTANT PARAGRAPH IN THE WHOLE PAPER

Define $\bar{T}_x = \bigcup_k \bar{T}_x^k$

Then $C^\infty(X)^\circ = \bigoplus_{x \in X} \bar{T}_x$

Define $\bar{T}^k X = \bigcup_{x \in X} \bar{T}_x^k$

Then $\bar{T}^k X$ forms a smooth vector bundle over X which is dual to the kth jet bundle over X.

The coalgebra $C^\infty(X)^\circ$ can be visualized schematically as below:

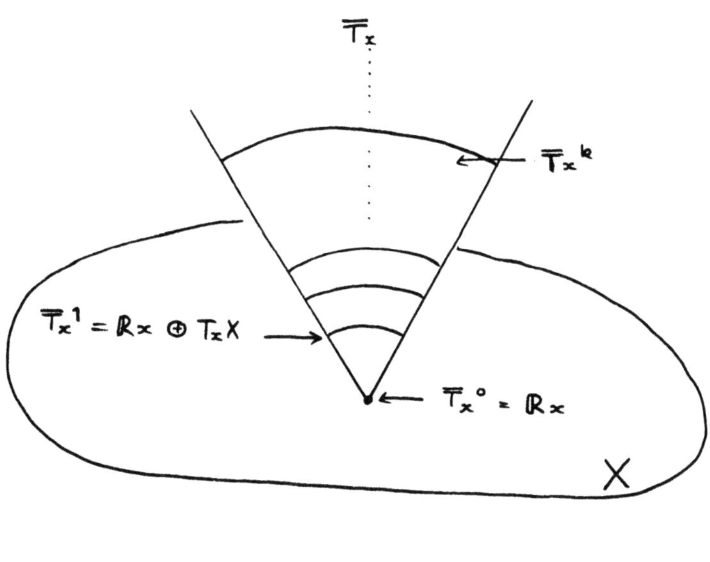

I.7. Picture of $C^\infty(X)^\circ$

The significance of the boxed statements is that they form a model which can be generalized. For smooth manifolds, the information in the box can be got directly from X. For graded manifolds (X,A), one forms $A(X)^{\circ}$ from the graded function ring $A(X)$ and interprets the information it contains by analogy with $C^{\infty}(X)^{\circ}$. For the manifold of smooth maps the technique is to postulate a reasonable generalization of the dual coalgebra, and, again, interpret the information it contains by analogy. The objects which generalize nicely and which are easy to work with are the total spaces $\overline{T}^{k}$, not the point set X.

The boxed statements are direct corollaries of the following propositions, which are further collections of comments on $C^{\infty}(X)^{\circ}$.

I.8 Proposition

1. If X is a nice manifold, and U is an open set in X, then $C^{\infty}(U)^{\circ} = \bigoplus_{u \in U} \overline{T}_u$.

Proof - sketch

This requires a little knowledge of coalgebras. The idea is as follows.

Since $C^{\infty}(U)$ is commutative, $C^{\infty}(U)^{\circ}$ is cocommutative, using a theorem in Sweedler's *Hopf Algebras* (8.0.5) [7] $C^{\infty}(U)^{\circ}$ can be written as a direct sum

$$C^{\infty}(U)^{\circ} = \bigoplus_{\alpha} C_{\alpha}$$

where each C_{α} is the maximal subcoalgebra containing a unique simple subcoalgebra, S_{α}. (A simple coalgebra is one which contains no proper non-zero subalgebras.) The idea is to identify C_{α} with $\overline{T}_u$ for some u in $C^{\alpha}(U)^{\circ}$.

Step 1. Show all S_{α} are one dimensional, $S_{\alpha} = \mathbb{R}\alpha$ where $\alpha : C^{\infty}(U) \to \mathbb{R}$ is an algebra homomorphism.

Step 2. Show any algebra homomorphism $\alpha : C^{\infty}(U) \to \mathbb{R}$ is given by $\alpha(f) = f(u)$ for some u in U.

Thus if α corresponds to the point u, $(I_u)^{\perp} \subset C_{\alpha}$. A further theorem of Sweedler (11.0.5) can be used to show that $\overline{T}_u = C_{\alpha}$.

Proof of Step 1. Notice that

$$S_{\alpha}^{\perp} = \{f \in C^{\infty}(U) : \langle S_{\alpha}, f \rangle = 0\}$$

is an ideal in $C^{\infty}(U)$. Show that $S_{\alpha}^{\perp}$ is maximal as follows. Suppose I is proper ideal containing $S_{\alpha}^{\perp}$. There is a surjective algebra homomorphism

$$\pi : C^{\infty}(U)/S_{\alpha}^{\perp} \to C^{\infty}(U)/I$$

inducing an injective coalgebra map

$$\pi' : (C^{\infty}(U)/I)' \to \underset{\substack{\parallel \\ S_{\alpha}}}{(C^{\infty}(U)/S_{\alpha}^{\perp})'} \subset C^{\infty}(U)^{\circ}.$$

Since S_α is minimal, π' must be an isomorphism. Hence $S_\alpha^\perp$ is maximal in $C^\infty(U)$.

Since $S_\alpha^\perp$ is maximal, $C^\infty(U)/S_\alpha^\perp$ is a simple commutative algebra over $\mathbb{R}$. This means $C^\infty(U)/S_\alpha^\perp$ is either $\mathbb{R}$ or $\mathbb{C}$. If $C^\infty(U)/S_\alpha^\perp$ were isomorphic to $\mathbb{C}$, we could find f in $C^\infty(U)$ such that $[f]^2 = -1$ in $C^\infty(U)/S_\alpha^\perp$, which would imply f^2+1 is in $S_\alpha^\perp$. But f^2+1 is invertible in $C^\infty(U)$, hence $S_\alpha^\perp = C^\infty(U)$, a contradiction.

Thus $C^\infty(U)/S_\alpha^\perp \cong \mathbb{R}$, and $S_\alpha = [C^\infty(U)/S_\alpha^\perp]' \subset C^\infty(U)^\circ$, hence we may write $S_\alpha = \mathbb{R}\alpha$ where $\langle\alpha,1\rangle = 1$. Since S_α is a subcoalgebra of $C^\infty(U)^\circ$, $\Delta\alpha = r\alpha\otimes\alpha$ for some r in $\mathbb{R}$. Since $\langle\alpha,1\rangle = \langle\Delta\alpha, 1\otimes 1\rangle$, r must be 1. Thus $\Delta\alpha = \alpha\otimes\alpha$, or, equivalently, α is a homomorphism.

Proof of Step 2

Choose a cover $\{U_\beta\}$ of U by coordinate neighbourhoods, with coordinate functions x_i^β, and a subordinate partition of unity Ψ_β.

Suppose $\overline{U}$ is compact. One may assume that there are finitely many β, and thus that the functions $\Psi_\beta x_i^\beta$ (no sum) provide coordinate functions of an embedding ρ of $\overline{U}$ in $\mathbb{R}^N$. Consider the function

$$F = \sum_{\beta,i} (\Psi_\beta x_i^\beta - \alpha(\Psi_\beta x_i^\beta))^2 .$$

If F has no solutions on U, F is invertible, and F is in ker α, hence $\ker\alpha = C^\infty(U)$: a contradiction. So F must have a solution. But $F(x) = 0$ if and only if $\Psi_\beta x_i^\beta(x) - \alpha(\Psi_\beta x_i^\beta) = 0$ for each (i,β). Thus the following diagram commutes

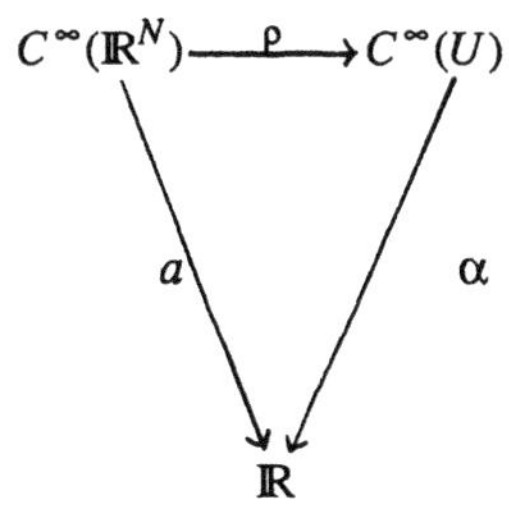

where a denotes evaluation at the point $a = (\alpha(\Psi_\beta x_i^\beta))$ in $\mathbb{R}^N$, and α is given by evaluation at $\rho^{-1}(a)$.

If $\overline{U}$ is not compact, the same argument will apply once I show $\alpha f = \alpha\Psi f$ for all f in U, for some Ψ with compact support.

To do this assume each U_β has compact closure, and that each u in U is contained in only finitely many U_β. Suppose $\alpha\Psi_\beta$ is non zero. Since $\Psi_\beta\Psi_\gamma = 0$ for all but finitely many Ψ_γ, $\alpha(\Psi_\beta\Psi_\gamma) = \alpha\Psi_\beta\alpha\Psi_\gamma$ implies that $\alpha\Psi_\gamma$ is zero for all but finitely many γ.

I want to show

$$\alpha\left(\sum_{\alpha\Psi_\gamma = 0} \Psi_\gamma\right) = 0 .$$

Then

$$1 = \alpha(1) = \alpha(\Sigma \Psi_\beta) = \alpha\left(\sum_{\alpha\Psi_\beta \neq 0} \Psi_\beta\right) + \alpha\left(\sum_{\alpha\Psi_\gamma = 0} \Psi_\gamma\right) = \sum_{\alpha\Psi_\beta \neq 0} \alpha\,\Psi_\beta .$$

Write $\sum_{\alpha\Psi_\beta \neq 0} \Psi_\beta = \Psi$. If this holds $\alpha(f) = \alpha(\Psi f)$ as desired.

Index the set $\{\Psi_\gamma : \alpha\Psi_\gamma = 0\}$ by the positive integers, and write $\phi_n = 1 - \sum_{i=1}^{n} \Psi_n - \Psi$. Thus $\phi_n(x) \geq 0$ and $\sum_{n>j} \phi_n = \chi_j$ is in $C^\infty(X)$. Schematically the function Ψ_n, Φ_n and χ_n can be graphed

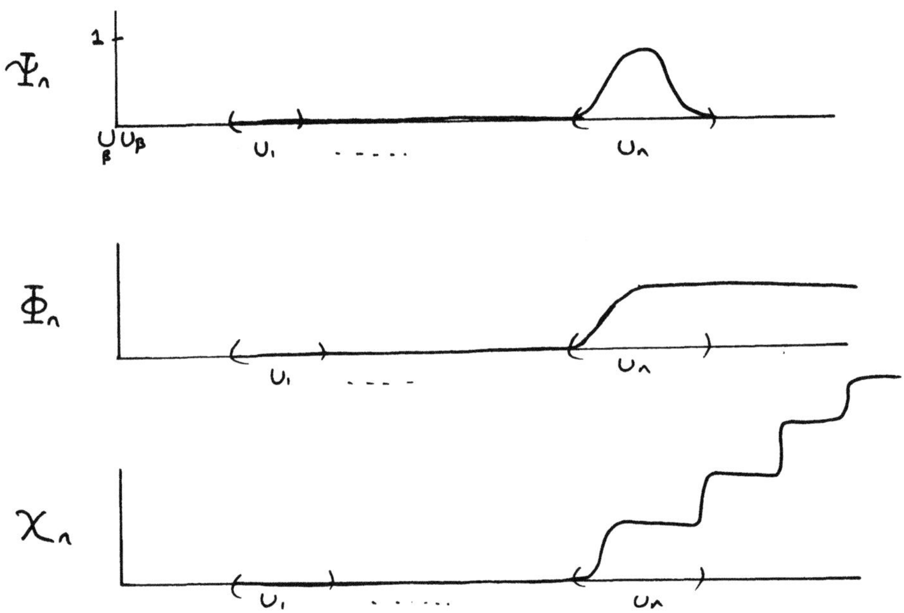

I.9 Lemma. If f in $C^\infty(X)$ has $f(x) \geq 0$, then $<\alpha, f> \geq 0$.

Proof. For $\varepsilon > 0$, $f + \varepsilon \geq \varepsilon$, and hence $\sqrt{f + \varepsilon}$ defines a smooth function on X. Thus

$$\alpha f + \varepsilon \;=\; \alpha(f + \varepsilon) \;=\; \left(\alpha(\sqrt{f + \varepsilon})\right)^2 \;\geq\; 0 .$$

Since $\alpha f + \varepsilon \geq 0$ for all ε, $\alpha f \geq 0$.

Since $1 - \Psi \geq 0$ $\Psi \geq 0$, $0 \leq \alpha\Psi \leq 1$. Suppose $\alpha(1 - \Psi) = r$ with $r < 1$. Then $\alpha(\phi_n) = \alpha(1 - \Psi) = r$. Again by the lemma $\alpha(\chi_n) \geq 0$. But writing $\chi_n = \phi_n + \phi_{n+1} + \cdots \phi_{n+k} + \chi_{n+k}$, thus $\alpha(\chi_n) = (k+1)r + \alpha\,\chi_{n+k}$. In particular, choosing k such that $\alpha(\Psi_1) < kr$,

$$\alpha(\Psi_k) \;=\; \alpha(\Psi_0) \;-\; kr \;<\; 0 .$$

This provides a contradiction, and the assumption that $1 - \alpha(\psi) \;=\; r < 1$ must be false. Hence $\alpha\psi = 1$ as desired.

Back to the main proof. So far we have

$$C^\infty(U)^\circ \;=\; \bigoplus_{u \in U} C_u$$

by identifying α with u in U. Moreover, $\bar{T}_u \subset C_u$, since $\bar{T}_u$ is irreducible with $\mathbb{R}u$ as the simple subcoalgebra.

One must show that $\bar{T}_u = C_u$. Again one can import coalgebraic techniques (Sweedler 11.0.5) to prove this. Essentially one has a filtration $C_u = \bigcup_k C_u^k$ got by saying $c \in C_u^k$ if and only if the expression

$$\Delta^k c \;=\; \sum_{(c)} c_{(1)} \otimes .. \otimes c_{(k+1)}$$

is such that every term $c_{(1)} \otimes \cdots \otimes c_{(k+1)}$ contains u as a factor. Then demonstrate that $C_u^k = \bar{T}_u^k$.

I.10 Proposition

For each x in X let E_x be a linear subspace of $\bar{T}_x$ such that E_x has finite dimension. Form $E = \coprod_x E_x$.

Define $\Gamma(U,E)$ to be the subset of all sections σ of E such that given any f in $C^\infty(U)$, the function

$$\sigma_f : U \;\rightarrow\; \mathbb{R}$$

given by

$$\sigma_f(u) = <\sigma(u),f>$$

is smooth.

Then

1) $\Gamma(\ ,E)$ is a sheaf of $C^\infty(\ )$-modules.

2) If X can be covered by open sets U_β such that

   a) $\Gamma(U_\beta,E)$ is a free $C^\infty(U_\beta)$ module on generators m_i^β

   b) $m_i^\beta(u)$ forms a basis of E_u for all u in U_β

   c)There exists f_β^i in $C^\infty(U_\beta)$ with $<f_\beta^j,m_i^\beta(u)> = \delta_{ij}$ for all u in U_β.

Then E can be given a unique structure as vector bundle over X such that $\Gamma(\ ,E)$ coincides with the sheaf of smooth sections of E.

Proof (Sketch)

The *bundle map*. Clearly we can define $\pi : E \to X$ by setting $\pi(e) = x$ if e is in E_x.

Topology on E. Give $C^\infty(X)^\circ$ a topology by defining a base of open sets at the origin

$$O(\varepsilon; a_1, \ldots, a_n) = \{\alpha \in C^\infty(X)^\circ : |<\alpha, a_i>| < \varepsilon \text{ for all } i\}.$$

Then use the injective map

$$E \to X \times C^\infty(X)^\circ$$

$$e \to (\pi(e), e)$$

to identify E with its image. The topology on E is the topology it has as a subspace of $X \times C^\infty(X)^\circ$.

Charts on E. I can assume that the open sets U_β are charts on X with smooth maps $\Psi_\beta : U_\beta \to \mathbb{R}^r$. An atlas for E can be constructed using the open sets $\pi^{-1}(U_\beta)$ with maps

$$\phi_\beta : \pi^{-1}(U_\beta) \to \mathbb{R}^r \times \mathbb{R}^s$$

$$\phi_\beta(e) = (\Psi_\beta(\pi(e)), <e,f_\beta^1>, \ldots, <e,f_\beta^s>).$$

These are the definitions one ought to check that ϕ_β is a homeomorphism onto an open set, that $\phi_\beta\phi_\alpha^{-1}$ is a diffeomorphism where defined, and that E is a smooth vector bundle with this topology. These details are tedious rather than tricky, and are omitted. See Batchelor [2] for details. (Caution: although the proof of this result is sound, the principal application in that paper is incorrect.)

The proofs of the boxed statements 1.5 now follow as corollaries of 1.7 and 1.9. The last statement that the bundles $\bar{T}^k$ are dual to the kth jet bundle is evident from the definition of the fibre $\bar{T}^k_x$.

I.11 Example: Differential operators and Lagrangians.

If E is a vector bundle over X, it is not hard to see that $\bar{T}^k \otimes E'$ (E' is the dual bundle of E) is dual to the kth jet bundle $J^k(E)$ of E. One may take symmetric powers of this bundle, replacing $\bar{T}^k_x \otimes E'_x$ with $S^h(\bar{T}^k_x \otimes E'_x)$. Summing over h (or over $h = N$ if you like finite dimensional fibres) gives a bundle

$$S^{*,k}(E) = \bigoplus_h S^{h,k}(E) = \bigoplus_h S^h(\bar{T}^k \otimes E').$$

Sections of $\bigoplus_k S^{*k}(E)$ give rise to k^{th} order polynomial differential operators. A section σ of $S^{*,1}(E)$ gives rise to a Lagrangian density

$$\mathcal{L}_\sigma : J^1(E) \to \mathbb{R}$$

as follows. Observe (see Abraham and Marsden [1], p.475)

$$\begin{aligned} J^1(E)_x &= E_x \oplus L(T_xX, E_x) \\ &= L(T_xX \oplus \mathbb{R}, E_x) \\ &= (T_xX \oplus \mathbb{R})' \otimes E_x \\ &= \left[\bar{T}^1_x \otimes E'_x\right]'. \end{aligned}$$

Thus a section s of $\bar{T}^1_x \otimes E'_x$ evaluates on elements j in $J^1(E)$ via

$$s(j) = \langle s(\pi(j)), j \rangle.$$

Any section σ of $S^{*1}(E)$ is a sum of products of sections like s above,

$$\sigma = \sum_k (s_1^k \cdots s_{L_k}^k)$$

hence

$$\mathcal{L}_\sigma(j) = \sum_k s_1^k(j) \cdots s_{L_k}^k(j).$$

To get an action integral, one needs a top form ω on X. The action integral is then

$$S(j) = \int_X \mathcal{L}_\sigma(j)\omega.$$

In this interpretation it is convenient to consider the Lagrangian density as a section of $S^{*1}(E) \otimes \Lambda^{\text{top}}X$. This is particularly true in the case where one has a symmetry group G acting on E and X, and hence on $S^{*2}(E) \otimes \Lambda^{\text{top}}X$. One can calculate the action of G on E , E', T_xX on T_xX', and hence on symmetric powers of these. The invariant subspace of $S^{*2}(E) \otimes \Lambda^{top} X$ can, in principle, be calculated.

For more on action integrals, see the section on functions on the space of smooth maps.

1.12 Geometric quantization

The observation that $\bar{T}X_x^1 = T_x X \oplus \mathbb{R}$ suggests that the geometric quantization programme can be interpreted as a recipe for lifting functions on X to sections of $\bar{T}_x^1$, or a map

$$Q : \Gamma(X, \bar{T}^0) \rightarrow \Gamma(X, \bar{T}^1).$$

The simplest prequantization schemes use a symplectic form ω to identify TX with T^*X, and a 1-form ν with a vector field ξ_ν. In this simplest scheme $Qf = f + \xi_{df}$, and Qf acts as operators on $C^\infty(X)$.

This interpretation in no way simplifies the technical difficulties of geometric quantization, but perhaps by interpreting functions as sections of $\bar{T}^0$ and the quantized operators as sections of $\bar{T}^1$ the appeal of this approach may be more apparent.

II Applications of dual coalgebras to graded manifolds

Graded manifolds were used by Kostant [6] to provide an object corresponding to Lie superalgebras as Lie groups correspond to Lie algebras. The idea is to specify the underlying 'body' manifold and the algebra of scalar superfields, and to require local θ-expansions of superfields.

II.1 Definition

A graded manifold is a pair (X,A) where X is a nice (= smooth, paracompact, Hausdorff, without boundary, but this last condition is not necessary) manifold, and A is a sheaf of $\mathbb{Z}_2$-graded commutative algebras such that

1) There is a map of sheaves of $\mathbb{Z}_2$-graded algebras

$$\varepsilon : A(\ ) \rightarrow C^\infty(\ ).$$

2) There is an open cover $\{U_i\}$ of X and isomorphisms

$$\tau_i : A(\ )|_{U_i} \rightarrow C^\infty(\ ) \otimes \Lambda <\theta_1^i, \ldots, \theta_s^i>|_{U_i}$$

of sheaves of $\mathbb{Z}_2$-graded algebras. Here $\Lambda <\theta_1^i, \ldots, \theta_s^i>$ is the exterior algebra generated by $\theta_1^i \cdots \theta_s^i$.

The standard examples are (X, C^∞), $(pt, \Lambda\mathbb{R}_s)$, (X, Ω) where Ω is the sheaf of smooth differential forms, and more generally $(X, \Gamma(\ , \Lambda E))$ where $\Gamma(\ , \Lambda E)$ is the sheaf of smooth sections of the exterior bundle of a smooth vector bundle E. This last example is typical. Given any graded manifold (X,A) there exists a bundle E such that the sheaf A is isomorphic to $\Gamma(\ , \Lambda E)$.

A map $f : (X,A) \rightarrow (Y,B)$ in the category of graded manifolds is an algebra homomorphism $f : B(Y) \rightarrow A(X)$.

The analysis of $A(X)^\circ$ proceeds in parallel with that of $C^\infty(X)^\circ$. Again define the ideal $I_x \subset A(X)$

$$I_x = \{a \in A(X) : \varepsilon a(x) = 0\}.$$

$$\bar{T}A_x^k = \left(I_x^{k+1}\right)^\perp = \left[A(X)/I_x^{k+1}\right]' \subset A(X)^\circ$$

II.2

Theorem - definitions - the structure of $A(X)^\circ$

Define $\bar{T}A_x = \bigcup_k \bar{T}A_x^k$

Then $A(X)^\circ = \bigoplus_{x \in X} \bar{T}A_x$

Define $\bar{T}A^k = \bigsqcup_{x \in X} \bar{T}A_x^k$

Then $\bar{T}A^k$ can be given the structure of a smooth vector bundle over X

The picture is also similar:

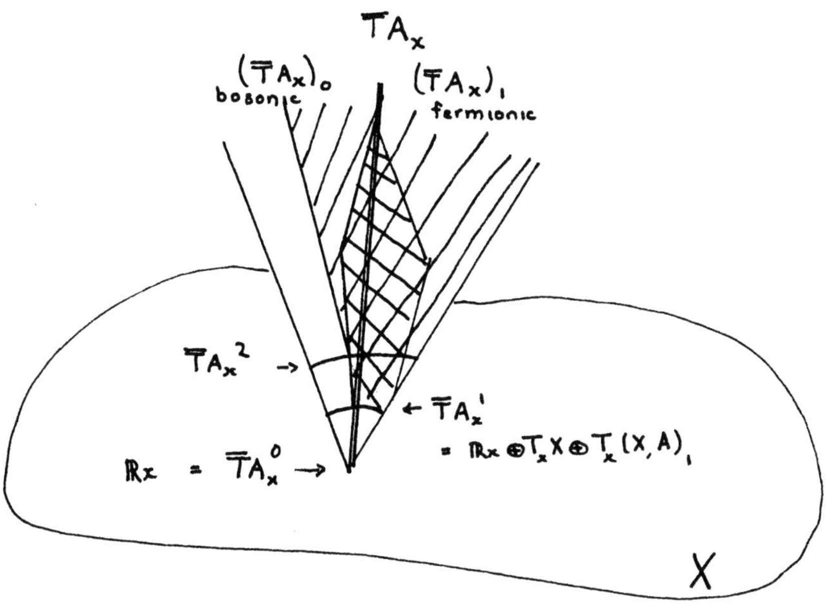

II.3 Picture of $A(X)^\circ$

Here $T_x(X,A)_1$ *is the subspace*

$$T_x(X,A)_1 \;=\; \{\alpha : A(x) \to \mathbb{R} : |\alpha| = 1$$

$$\alpha(ab) \;=\; \alpha(a)\varepsilon b(x) + (-1)^{|a|}\varepsilon a(x)\alpha(b)\}.$$

If A *is isomorphic to* $\Gamma(\ ,\Lambda E)$, $T_x(X,A)_1$ *is the dual vector bundle of* E *(see II.4 below). The interpretion of the shading is as follows:*

Unshaded sector : differentiations of bosonic type, e.g.

$$\frac{\partial^{i_1}}{\partial x_1} \cdots \frac{\partial^{i_r}}{\partial x_r}\bigg|_x .$$

This is a canonical subcoalgebra of $\overline{T}A_x$.

 : differentiations of fermionic type, e.g.

elements $\frac{\partial^{j_1}}{\partial \theta_1} \cdots \frac{\partial^{j_s}}{\partial \theta_s}\bigg|_x$

 : elements of mixed type, e.g.

$$\frac{\partial^{i_1}}{\partial x_1} \cdots \frac{\partial^{i_r}}{\partial x_r}\frac{\partial^{j_1}}{\partial \theta_1} \cdots \frac{\partial^{j_s}}{\partial \theta_s}$$

In order to recognize subspaces of the last two types one must effectively choose a local coordinate system.

There are two observations. First, the elements in $T_x(X\ ,A)_1$, or in any of the shaded area, cannot be represented by equivalence classes of trajectories in X. In the case of $C^\infty(X)$, all the information encoded in $C^\infty(X)^\circ$ could be derived from X. By contrast, $A(X)^\circ$ contains information encoded in *A(X)* which cannot be derived from X. One must accept this and deal with $A(X)^\circ$.

Secondly, $A(X)^\circ$ plays the role of a generalized point set, in that $A(X)$ is included as a subalgebra of $(A(X)^\circ)'$ (see prop. II.4 below). Thus elements of *A(X)* are distinct if and only if there is a point in $A(X)^\circ$ on which they take different values. For $C^\infty(X)$, X itself 'separates' elements of $C^\infty(X)$ in this way; for *A(X)*, X fails to separate elements and one must use $A(X)^\circ$ (or a suitable subset at $A(X)^\circ$) if one wants a 'point set' which separates elements of *A(X)*. In II.8 below $\overline{T}A^2$ is used to build a supermanifold corresponding to *(X,A)*.

Some useful properties of $A(X)^\circ$ are summarized in the following proposition. The statements in the box will follow as corollaries.

II.4 Proposition

1. For an open set U in X, $A(U)^\circ = \bigoplus_{u \in U} \overline{T}A_u$.

2. If U is an open set in $\mathbb{R}^r$

$$(C^\infty(U)\otimes\Lambda<\theta_1,\ldots,\theta_s>)^\circ = \left< \frac{\partial^{\Sigma i_l + j_k}}{\partial x_1^{i_1}\cdots\partial x_r^{i_r}\,\partial\theta_1^{j_1}\cdots\partial\theta_s^{j_s}} \;\middle|\; \begin{array}{l} i_n = 0, 1, 2\cdots \\ : j_k\in\{0,1\} \\ u\in U \end{array} \right>$$

Comultiplication is given by

$$\Delta \left.\frac{\partial^{\Sigma i_r + j_k}}{\partial_{x_1}^{i_1}\cdots\partial x_r^{i_r}\,\partial\theta_1^{j_1}\cdots\partial\theta_s^{j_s}}\right| = \Sigma\left[c(\underline{i})\sigma(j)\frac{\partial^{\Sigma i'_k + j'_k}}{\partial x_1^{i'_1}\cdots\partial x_r^{i'_r}\,\partial\theta_1^{j'_i}\cdots\partial\theta_s^{j'_s}} \right.$$

$$\left.\otimes\ \frac{\partial^{\Sigma i''_n + j''_k}}{\partial x_1^{i''_1}\cdots\partial x_r^{i''_r}\,\partial\theta_1^{j''_1}\cdots\partial\theta_s^{j''_s}}\right]$$

$$i_h = i'_h + i''_h\,,\quad jk = j'_n + j''_k\,,\quad c(\underline{i}) = \prod_j \binom{i_n}{i'_n}$$

and the sign $\sigma(j)$ is chosen to make the following equation hold

$$\frac{\partial^{\Sigma_{jn}}}{\partial\theta_1^{j_1}\cdots\partial\theta_s^{j_s}} = \sigma(j)\frac{\partial^{\Sigma_{j'k}}}{\partial\theta_1^{j'_1}\cdots\partial\theta_s^{j'_s}}\ \frac{\partial^{\Sigma_{j''_n}}}{\partial\theta_1^{j''_1}\cdots\partial\theta_s^{j''_s}}]$$

3. The full linear dual of $A(X)^\circ$, $(A(X)^\circ)'$, is an algebra and $A(X)$ is included as a subalgebra of $(A(X)^\circ)'$.

4. The augmentation $\varepsilon : A(X)\to C^\infty(X)$ induces an incusion $\varepsilon^\circ : C^\infty(X)^\circ \to A(X)^\circ$. (In the picture, the image of $\bar{T}_x$ is the unhatched area of $\bar{T}A_x$.)

5. Let F_x be a finite dimensional subspace of $\bar{T}A_x$. Then $\coprod_x F_x = F$ can be given the structure of a smooth vector bundle over X, if F satisfies the same properties required of E in prop. I.4. In particular $\bar{T}A^k$ is a vector bundle, and if $A(X)=\Gamma(X,\Lambda E)$ for some bundle E over X, $\bar{T}A^1 = \mathbb{R}\oplus(TX\oplus E')$.

Proofs.

1. Sketch. First I need a lemma about coalgebras.

II.5 Lemma

A graded cocommutative coalgebra is the direct sum of its irreducible components. See Sweedler 8.0.5 for the proof. Irreducible subcoalgebras contain unique simple subcoalgebras. Thus an irreducible component can be labelled by the simple subcoalgebra it contains.

Since any maximal ideal in $A(U)$ contains ker ε, all simple subcoalgebras are in the image of $\varepsilon^\circ : C^\infty(U)^\circ \to A(U)^\circ$. Thus

$$A(U)^\circ = \bigoplus_{u\in U} S_u$$

where S_u is the irreducible component containing $\mathbb{R}u$. It is not hard to see that $S_u > \bar{T}A_u$. As in I.7 the result now follows essentially from Sweedler 11.0.5.

2. This is a direct corollary of 1.

3. Multiplication in $(A(X)^\circ)'$ is given by

$$<\alpha, a_1 a_2> \;=\; \sum_{(\alpha)} (-1)^{|\alpha_{(2)}||a_{(1)}|} <\alpha_{(1)}, a_1> <\alpha_{(2)}, a_2>$$

for a_1, a_2 in $(A(X)^\circ)'$. the natural map

$$A(X) \;\to\; (A(X)^\circ)'$$

will always be a homomorphism. The fact that this map is injective follows from the fact that a non-zero a in $A(X)$ must have a non-zero restriction to some coordinate patch $A(U) \cong C^\infty(U) \otimes \Lambda \mathbb{R}^s$, and part 2.

4. This follows from I.2, section 4.

5. This is essentially the same as I.9.

Graded manifolds were used by Kostant for the purpose of studying graded Lie groups, and dual coalgebras play a central role in this thoery.

II.6 Definition.

A graded Lie group is a graded manifold (G, A) such that the dual coalgebra $A(G)^\circ$ is a Hopf algebra, that is $A(G)^\circ$ has a multiplication

$$m : A(G)^\circ \otimes A(G)^\circ \;\to\; A(G)^\circ$$

making $A(G)^\circ$ an associative algebra and an antipodal map $s : A(G)^\circ \to A(G)^\circ$ such that the following identities hold.

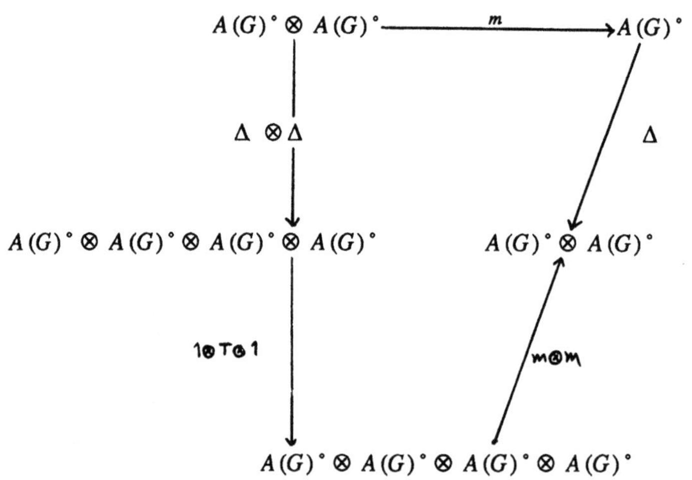

(1)

where $T(\alpha \otimes \beta) = (-1)^{|\alpha|\,|\beta|} \beta \otimes \alpha$.

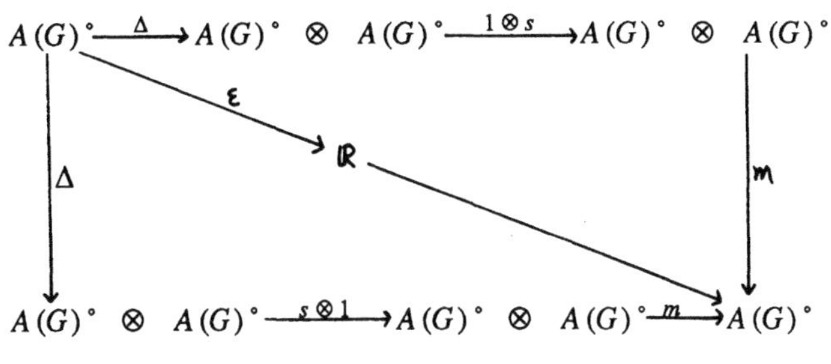

(2)

II.7 Example

Let $\mathcal{G} = \mathcal{G}_0 \oplus \mathcal{G}_1$ be any Lie superalgebra, and let G be any Lie group with Lie algebra $\mathcal{G}_1$, and assume that the action of $\mathcal{G}_0$ on $\mathcal{G}_1$ exponentiates to an action of G on $\mathcal{G}_1$. Write $U(\mathcal{G}_0 \oplus \mathcal{G}_1)$ for the universal enveloping algebra.

Then

$(G, C^\infty(\ ) \otimes \Lambda(\mathcal{G}_1'))$ is a graded Lie group.

If (G,A) is a graded Lie group, then

$A(G)^\circ \cong \mathbb{R}G \otimes U(\mathcal{G}_0 \oplus \mathcal{G}_1)$

where $\mathcal{G}_0 \oplus \mathcal{G}_1$ is a Lie superalgebra with $\mathcal{G}_0$ the Lie algebra at G, and $A \cong C^\infty(\ ) \otimes \Lambda(\mathcal{G}_1')$.

As a final example of the use of dual coalgebras, I want to use $A(X)^\circ$ to construct a supermanifold corresponding to (X,A).

II.8 QUICK REVIEW OF SUPERMANIFOLD THEORY

The idea is to mimic the ordinary construction of graded manifold with a suitable graded commutative ring replacing the real numbers. More explicitly, the constructions are as follows.

Replace $\mathbb{R}$ *by* $B_L = \Lambda\mathbb{R}^L$. Take B_L to be the algebra generated by anti-commuting elements $\gamma_1, \ldots, \gamma_L$, where each γ_i is to be thought of as a distinct square root of 0. As usual B_L is graded with B_{L_0} (B_{L_1}) being sums of products of an even (odd) number of the γ_j. The number L may be finite or infinite.

Replace $\mathbb{R}^n$ *by* $E^{r,s}$. Set

$$E_L^{r,s} = \left\{ (u_1, \ldots, u_r, v_1, \ldots, v_s) : u_i \in B_{L_0}, v_j \in B_{L_1} \right\}.$$

The topology in $E_L^{r,s}$ is usually (but not always) taken to be the topology of $E_L^{r,s}$ considered as a finite dimensional real vector space. The subscript L will usually be suppressed.

Smooth functions on $E^{r,s}$. There is again some choice but usually the algebra of smooth functions is that generated by the following subalgebras:

a) Constant B_L-valued functions.

b) The subalgebra generat ed by the projections

$$\theta_j : E^{r,s} \to B_{L1},\ \theta_j(u_1, \ldots, u_r, v_1, \ldots, v_s) = v_j .$$

The subalgebra generated by the 'expansions' of smooth functions f in $C^\infty(\mathbb{R}^r)$. Using the map

$$\varepsilon : B_L \to \mathbb{R}^r$$

given by setting all r_i to 0, one can expand f using Taylor's formula to a function $\check{f}$

$$\check{f}(u_1, \ldots, u_r, v_1, \ldots, v_s) =$$

$$+ \sum_{\underline{i}} b(\underline{i})(\partial(\underline{i})f)(\varepsilon(u_1), \ldots, \varepsilon(u_r))\, p(\underline{i})(u_1 - \varepsilon(u_1), \ldots, u_r - \varepsilon(u_r)) \qquad \dagger$$

where

$$\underline{i} = (i_1, \ldots, i_r) \text{ is an } r\text{-tuple of non-negative integers}$$

$$\partial(\underline{i}) = \frac{\partial^{\sum i_k}}{\partial x_1^{i_1} \cdots \partial x_r^{i_r}}$$

$$b(\underline{i}) = {}^1/i_1! \cdots i_r!$$

$$p(\underline{i})(\omega_1, \ldots, \omega_r) = \omega_1^{i_1} \cdots \omega_r^{i_r} \qquad \omega_i \in B_{L0} .$$

There are no convergence problems since each element $u_i - \varepsilon(u_i)$ is nilpotent.

With this information one can define maps between subsets U, U' of $E^{r,s}$ to be supersmooth if the projections on each coordinate is in $S(E^{r,s})$. Supermanifolds can then be defined as topological spaces which are locally homeomorphic to $E^{r,s}$ in such a way that the homeomorphisms of subsets of $E^{r,s}$ arising from comparison of two local charts are supersmooth.

There is a method of starting with a graded manifold (X,A) and producing a supermanifold S_A by setting

$$S_A = \{\alpha : A(X) \to B : \alpha \text{ is an algebra homomorphism}\} .$$

It can be shown that this assignment establishes an equivalence between the category of graded manifolds and one particular category of supermanifold.

II.9 Direct construction of S_A.

The bundle $T(X,A)_1$ can be used to give a direct method of construction S_A. First construct $\tilde{X}$ by 'fattening' X as follows. Choose coordinate charts $\{(U_\alpha, \psi_\alpha)\}$ on X, with $\psi_\alpha : U_\alpha \to \mathbb{R}^r$, and write $U_{\alpha,\beta} = \psi_\alpha(U_\alpha \cap U_\beta)$, and $\tau_{\beta\alpha} = \psi_\beta \psi_\alpha^{-1} : U_{\alpha,\beta} \to U_{\beta,\alpha}$. Then X can be given by

$$X = \coprod_{\alpha} U_\alpha / \sim$$

where $u_\alpha \sim u_\beta$ is and only if $\tau_{\beta\alpha}\, u_\alpha = u_\beta$.

The maps $\tau_{\beta\alpha}$ can be expanded by the formula † to give maps

$$\tilde{\tau}_{\alpha\beta} \,:\, \varepsilon^{-1}(U_{\alpha,\beta}) \rightarrow \varepsilon^{-1}(U_{\beta,\alpha})$$

where ε is considered as a map $\varepsilon : E^{r,0} \rightarrow \mathbb{R}^r$. These transition functions enable one to define

$$\tilde{X} = \coprod_{\alpha} \varepsilon^{-1}(U_\alpha)/_{\approx}$$

where $\omega_\alpha \approx \omega_\beta$, if $\tilde{\tau}_{\alpha\beta}\, \omega_\alpha = \omega_\beta$ for ω_α in $\varepsilon^{-1}(U_{\alpha,\beta})$. There is clearly a map $\tilde{X} \xrightarrow{\varepsilon} X$, and $\tilde{X}$ is a super-manifold of dimension *(r,0)*.

Secondly construct the pull-back $\tilde{T}(X,A)_1$

$$\begin{array}{ccc} \tilde{T}(X,A)_1 & \xrightarrow{\varepsilon_*} & T(X,A)_1 \\ \downarrow & & \downarrow \\ \tilde{X} & \xrightarrow{\varepsilon} & X \end{array}$$

This gives a finite dimensional real vector bundle over our 'fat' manifold $\tilde{X}$.

Now replace each fibre $\tilde{T}(X,A)_{1\tilde{x}}$ by $\tilde{T}(X,A)_{1\tilde{x}} \otimes B_1$ to form the bundle $\tilde{T}(X,A)_1 \otimes B_1$. It can be verified that $\tilde{T}(X,A)_1 \otimes B_1 = S_A$.

III The Dual Coalgebra of the Manifold of Smooth Maps

Let *X,Y* be nice manifolds, and let *M(X,Y)* = *M* denote the set of smooth maps from *X* to *Y*. The aim is to get a working model for $C^\infty(M)^\circ$, without specifying $C^\infty(M)$ or even the smooth structure on *M* precisely. The differential information can be obtained, and various candidates for $C^\infty(M)$ postulated.

The idea

Treat $\mathbb{R}$ as $C^\infty(*)$ where $*$ is a point. In this manner, *Y* may be identified with *M(*,Y)* and $C^\infty(Y)^\circ \subset \mathrm{Hom}(C^\infty(Y), \mathbb{R})$ may be identified with $C^\infty(M(*,Y))^\circ$. Thus to study $C^\infty(M(X,Y))^\circ$ one should find a coalgebra *P* with a map

$$\pi : P \rightarrow \mathrm{Hom}(C^\infty(Y), C^\infty(X))$$

such that the comultiplication on *P* is 'dual' to the multiplication in $C^\infty(Y)$ and $C^\infty(X)$ in that

$$<\pi p, fg> = \sum_{(p)} <\pi p_{(1)}, f> <\pi p_{(2)}, g>$$

where $\Delta p = \sum_{(p)} p_{(1)} \otimes p_{(2)}$.

This condition can be described in another way. Suppose we have $\pi : P \to \mathrm{Hom}(C^\infty(Y), C^\infty(X))$. The linear map π provides a linear map (also denoted π),

$$\pi : C^\infty(Y) \to \mathrm{Hom}(P, C^\infty(X)).$$

Now $\mathrm{Hom}(P, C^\infty(X))$ has an algebra structure. Given α, β in $\mathrm{Hom}(P, C^\infty(X))$, p in P define

$$<p, \alpha\beta> = \sum_{(p)} <p_{(1)}\alpha> <p_{(2)}\beta>.$$

The unit is given by sending p to the constant $\varepsilon(p)$ in $C^\infty(X)$.

III.1 Definition

If P is a coalgebra, and A,B are algebras say

$$\pi : P \to \mathrm{Hom}(A,B)$$

measures (A,B) if

$$\pi : A \to \mathrm{Hom}(P,B)$$

is an algebra homomorphism. P is called a *measuring* coalgebra.

III.2 Lemma. The map $\pi : P \to \mathrm{Hom}(A,B)$ measures (A,B) if and only if

1) $<\pi p, q_1 q_2> = \sum_{(p)} <\pi(p_{(1)}), q_1> <\pi(p_{(2)}), q_2>$

and

2) $<\pi p, 1> = \varepsilon(p)$

where $\varepsilon : P \to \mathbb{R}$ is the counit in the coalgebra P.

Proof By definition $<\pi p, a> = <p, \pi a>$. Thus

$$<p, \pi(a_1 a_2)> = <\pi p, a_1 a_2>$$

and

$$<p, \pi(a_1)\pi(a_2)> = \sum_{(p)} <p_{(1)}, \pi(a_1)> <p_{(2)}, \pi(a_2)>$$

$$= \sum_{(p)} <\pi(p_{(1)}), a_1> <\pi(p_{(2)}), a_2>.$$

Thus it is evident that $\pi(a_1)\pi(a_2) = \pi(a_1 a_2)$ if and only if formula 1 holds.

Similarly, since $<\pi p, 1> = <p, \pi(1)>$ and the identity in $\mathrm{Hom}(P,B)$ is the counit of P followed by the unit in B, $P \xrightarrow{\varepsilon} \mathbb{R} \to B$, $\pi(1)$ is the identity in $\mathrm{Hom}(P,B)$ if and only if fomula 2 holds.

III.3 Example

The inclusion $C^\infty(X)^\circ \to \mathrm{Hom}(C^\infty(X), \mathbb{R})$ measures $(C^\infty(X), \mathbb{R})$. The verifications are the same as those in Proposition 1.1, part 4.

A measuring coalgebra $\pi : P \to \mathrm{Hom}(A,B)$ is universal if for any other coalgebra C with $\psi : C \to \mathrm{Hom}(A,B)$ measuring *(A,B)* there is a unique coalgebra map $\eta : C \to P$ making the following diagram commute.

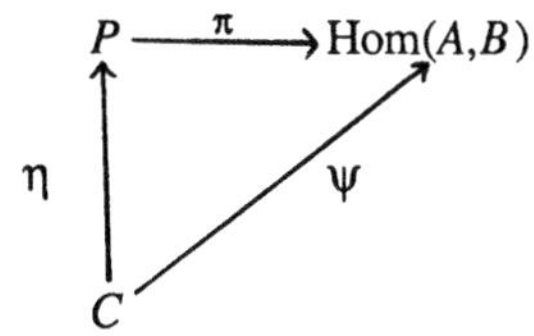

III.4 Proposition

There exists a unique universal measuring coalgebra $\tilde{\pi} : \tilde{P} \to \mathrm{Hom}(A,B)$.

Proof. Sketch (after Sweedler).

The first step is to introduce a cofree coalgebra on V, $\pi : C \to V$, the analogue of the tensor algebra. Here V is a vector space and C is a coalgebra with a fixed map π such that if D is any other coalgebra, and $\psi : D \to V$ is linear, there is a unique coalgebra map $p : D \to C$ making the following diagram commute

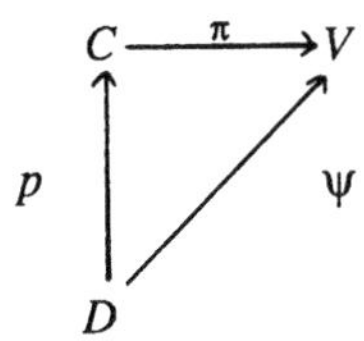

The cofree coalgebra is constructed as follows. Take the tensor algebra on the full linear dual of V, $T(V')$, and take its dual coalgebra $T(V')^\circ$. The inclusion $i : V' \to T(V')$ gives a map

$$i^\circ : T(V') \to V'' .$$

If V were finite, $T(V')^\circ \to V'' = V$ would have the desired properties. If V is not finite, define C to be the sum of all subcoalgebras of $T(V')^\circ$ whose image under i° lies in $V \subset V''$. One must verify that this object satisfies the conditions.

To get the measuring coalgebra, construct the cofree coalgebra on $\mathrm{Hom}(A,B)$, $\pi : D \to \mathrm{Hom}(A,B)$. Restrict π to the sum $\Sigma E = \tilde{P}$ of all subcoalgebras of D for which the map $\pi : E \to \mathrm{Hom}(A,B)$ actually measures.

So far the existence of a measuring coalgebra $\tilde{\pi} : \tilde{P} \to \mathrm{Hom}(C^\infty(Y), C^\infty(X))$ is established, but there is no reason to suppose that $\tilde{P}$ will have the desirable properties that $C^\infty(Y)^\circ$ enjoys. To ensure that P decomposes as a direct sum of pointed (graded) cocommutative subcoalgebras, $\tilde{P}$ must be suitably restricted.

III.5 Definition - Proposition

Let *(X,A)*, *(Y,B)* be graded manifolds and let *P(B,A)* be the sum of all pointed graded cocommutative subalgebras of the universal measuring coalgebra $\tilde{P}$ of *(B,A)*. Then if C is a pointed graded cocommutative coalgebra and $\psi : C \to \mathrm{Hom}(B,A)$ measures *(B,A)*, the universal map $\eta : C \to \tilde{P}$ has its image in *P(B,A)*. The coalgebra *P(B,A)* will be called the mapping coalgebra for *(B,A)*.

Proof

Let C be pointed graded cocommutative and suppose $\psi : C \to \mathrm{Hom}(B,A)$ measures. The universal mapping property guarantees a map $\eta : C \to \tilde{P}(B,A)$. The image of C will form a subcoalgebra E of $\tilde{P}(B,A)$. It is enough to show that E must be pointed graded cocommutative.

Let S be a simple subcoalgebra in E. Thus $\eta^{-1}S$ is a subcoalgebra of C, containing (at least) one simple subcoalgebra $D = \mathbb{R}d$, with $\Delta d = d \otimes d$, and $\varepsilon(d) = 1$. Since η is a coalgebra map, $\varepsilon(d) = \varepsilon(\eta(d))$. This implies $\eta(d)$ is non-zero. Since $\mathbb{R}\eta(d)$ spans a subcoalgebra of S, and S is simple, it must be that $S = \mathbb{R}\eta(d)$. This proves E is pointed.

For graded cocommutativity, I need to check that for e in E

$$\Delta e = \sum_{(e)} e_{(1)} \otimes e_{(2)} = \sum_{(e)} (-1)^{|e_{(1)}||e_{(2)}|} e_{(2)} \otimes e_{(1)} .$$

Take c in C with $\eta c = e$. The result follows since

$$\Delta e = \Delta\eta c = \sum_{(c)} \eta c_{(1)} \otimes \eta c_{(2)}$$

$$= \sum_{(c)} (-1)^{|c_{(1)}||c_{(2)}|} \eta c_{(2)} \otimes \eta c_{(1)} .$$

(Coalgebra morphisms are required to preserve Z_2 grading.)

Our structure theorems of the previous sections generalize.

II.6 Lemma.

1) Let $\sigma : B(X) \to A(Y)$ be an algebra homomorphism. Then $\mathbb{R}\sigma$ is a coalgebra with comultiplication $\sigma \to \sigma \otimes \sigma$, $\varepsilon(\sigma) = 1$ and

$$\mathbb{R}\sigma \to \mathrm{Hom}(B(X), A(Y))$$

measures.

2) If $\psi : C \to \mathrm{Hom}(B(X), A(Y))$ is any pointed graded cocommutative measuring coalgebra, and ψ is injective, then $\eta : C \to P(B,A)$ is injective.

The proof of 1 is an exercise in the definition of a measuring coalgebra. Part 2 is immediate.

If $\sigma : B(Y) \to A(X)$ is an algebra homomorphism, let I_σ denote the kernel of σ. We can define $\bar{T}(B,A)^k_\sigma$ to be the subcoalgebra of $P(B,A)$ given by

$$\bar{T}(B,A)^k_\sigma = \Big\{ c \in P(B,A) : \Delta^k c = \sum_{(c)} c_{(1)} \otimes \cdots \otimes c_{(k+1)}$$

$$\text{each term } c_{(1)} \otimes \cdots \otimes c_{(k+1)} \text{ contains}$$

$$\sigma \text{ as a factor} \Big\}.$$

III.7

STRUCTURE THEOREM FOR P(B,A)

$$P(B,A) = \bigoplus_{\sigma} \bar{T}(B,A)_\sigma$$

where the sum is taken over all algebra homorphisms $\sigma : B(Y) \to A(X)$
and

$$\bar{T}(B,A)_\sigma = \bigcup_k \bar{T}(B,A)^k_\sigma$$

Also $\bar{T}(B,A)^k_\sigma$ has the property that the following diagram commutes

$$\begin{array}{ccc} \bar{T}(B,A)^k_\sigma & \xrightarrow{\pi} & \mathrm{Hom}(B(Y), A(X)) \\ {\scriptstyle\bar{\pi}} \downarrow & \nearrow & \\ \mathrm{Hom}(B(X)/I^{k+1}_\sigma, A(X)) & & \end{array}$$

and the map $\bar{\pi}$ measures $\mathrm{Hom}(B(Y)/I^{k+1}_\sigma, A(X))$.

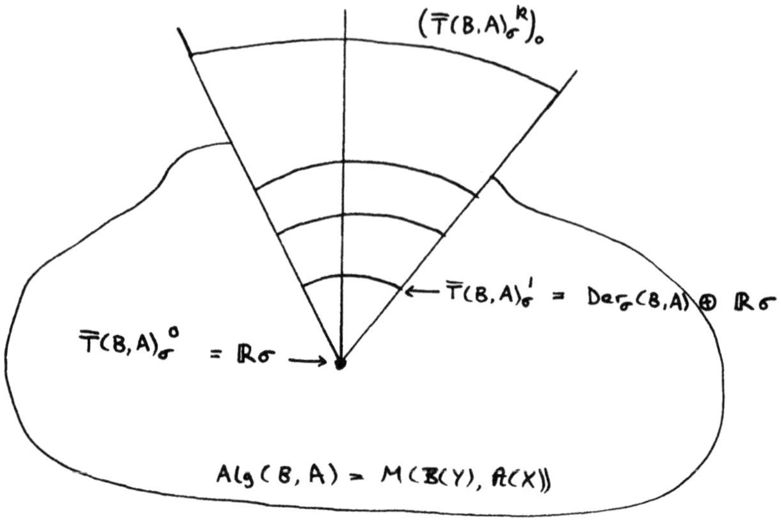

III.8 Picture for $P(B,A)$.

This space, $P(B,A)$, is to be interpreted as the dual coalgebra of the graded manifold of smooth maps from (Y,A) to (X,B). Compare this picture with that of $A(X)^\circ$(II.3). Here $\mathrm{Der}_\sigma(B,A)$ is the set of derivations with respect to σ:

$$\mathrm{Der}_\sigma(B,A) = \{r : B \to A \ : r(b_1b_2) = r(b_1)\sigma(b_2) + (-1)^{|r||b_1|}\sigma(b_2)r(b_1)\} \ .$$

The intention now is to study the diffential topology of this 'graded manifold' by studying $P(B,A)$. The following example, taking (Y,A), (X,B) to be (Y,C^∞), (X,C^∞) respectively, demonstrates the geometric nature of this construction.

III.9 Examples

1. Let $q : X \to Y$ be a smooth map between nice manifolds. Then the associated algebra map

$$q : C^\infty(Y) \to C^\infty(X)$$

is a (group-like) element of $P(C^\infty(Y), C^\infty(X))$. Conversely, given a group-like element p in $P(C^\infty(X), C^\infty(Y))$, πp in $\mathrm{Hom}(C^\infty(X), C^\infty(Y))$ is an algebra homomorphism, and consequently πp corresponds to a unique smooth map $\pi p : X \to Y$.

2. Again let $q : X \to Y$ be a smooth map. Suppose $\phi : X \times (-\varepsilon, \varepsilon) \to Y$ is a smooth map such that $\phi(x, 0) = q(x)$. Then ϕ determines a linear map

$$[\phi] : C^{\infty}(Y) \rightarrow C^{\infty}(X)$$

given by

$$[\phi]f(x) = \frac{d}{dt} f(\phi(x,t))\Big|_{t=0}$$

for f in $C^{\infty}(Y)$. It can be verified that the space $C = \mathbb{R}q \oplus \{[\phi] : \phi \text{ as above}\}$ forms a coalgebra with $\Delta q = q \otimes q$, $\Delta[\phi] = [\phi] \otimes q + q \otimes [\phi]$. The obvious inclusion $C \rightarrow \mathrm{Hom}(C^{\infty}(Y), C^{\infty}(X))$ measures; hence C can be considered as a subcoalgebra of $P(C^{\infty}(Y), C^{\infty}(X))$.

The construction in 2 is an obvious generalization of the space of equivalence classes of curves through q: thus this geometric description of tangent space is included in the algebraic description.

3. Given an element p in $\bar{T}(C^{\infty}(Y), C^{\infty}(X))^k_{\sigma}$ we get a lift of $\sigma : X \rightarrow Y$ to $p : X \rightarrow \bar{T}Y^k$

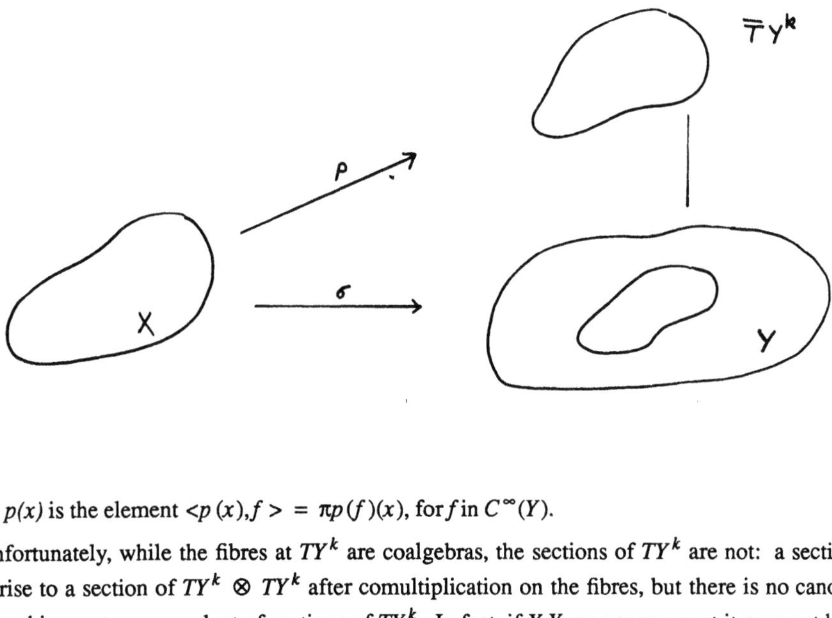

where $p(x)$ is the element $\langle p(x), f \rangle = \pi p(f)(x)$, for f in $C^{\infty}(Y)$.

Unfortunately, while the fibres at TY^k are coalgebras, the sections of TY^k are not: a section of TY^k gives rise to a section of $TY^k \otimes TY^k$ after comultiplication on the fibres, but there is no canonical way to write this as a tensor product of sections of TY^k. In fact, if X,Y are noncompact it may not be possible to write this as a tensor product of sections of TY^k. For this reason, sections of TY^k do not, in general, give rise to elements of $P(C^{\infty}(Y), C^{\infty}(X))$.

It is helpful to have a working description of $\bar{T}(B,A)_{\sigma}$, and such a description can be given in terms of the derivations of A and B. Write $\mathrm{Der}(A)$ ($\mathrm{Der}(B)$) for the Lie superalgebra of derivations of $A(B)$ respectively. Let $U(\mathrm{Der}(A))$, $U(\mathrm{Der}(B))$ denote the respective universal enveloping algebras. As such, $U(\mathrm{Der}(A))$, $U(\mathrm{Der}(B))$ have the structure of Hopf algebras, and, in particular, $U(\mathrm{Der}(A)) \otimes U(\mathrm{Der}(B)) = U(A,B)$ has the structure of a coalgebra.

III.10 Proposition

For each algebra homomorphism $\sigma : B \to A$ there is a measuring map

$$\phi_\sigma : U(A,B) \to \mathrm{Hom}(B,A)$$

given by

$$\phi_\sigma(\alpha_1 \cdots \alpha_r \beta_1 \cdots \beta_s)(b) = \alpha_1\alpha_2 \cdots \alpha_r \,\sigma\, \beta_1 \cdots \beta_s(b)$$

for β_i in Der B, α_j in Der (A).

Thus there is a coalgebra map

$$p_\sigma : U(A,B) \to P(B,A)$$

such that $\phi_\sigma = \pi p_\sigma$. Moreover, the image of p_σ lies in $\bar{T}(B,A)_\sigma$.

Proof

The proofs are easy, once one is given explicitly the comultiplication and counit on $U(A,B)$. These are defined as follows:

$$\Delta(\alpha_1 \cdots \alpha_r \beta_1 \cdots \beta_s)$$
$$= \sum \sigma(\underline{i},\underline{j})\alpha_{i_1} \cdots \alpha_{i_k} \beta_{j_1} \cdots \beta_{j_h} \otimes \alpha_{i'_1} \cdots \alpha_{i'_{k'}} \beta_{j'_1} \cdots \beta_{j'_{h'}}$$

where $i_\alpha < i_{\alpha+1}$, etc. $\{i_1 \cdots i_k\} \cup \{i'_1 \cdots i'_{k'}\}$ is a partition of the set $\{1, ..., r\}$, and $\sigma(\underline{i},\underline{j})$ is an apropriate sign. Thus $\Delta\alpha = \alpha \otimes 1 + 1 \otimes \alpha$ for α in Der A (or Der B), $\Delta\alpha\beta = \alpha\beta \otimes 1 + \alpha \otimes \beta + (-1)^{|\alpha||\beta|}\beta \otimes \alpha + 1 \otimes \alpha\beta$ for α, β in Der A $\oplus$ Der B.

The counit $\varepsilon : U(A,B) \to \mathbb{R}$ is given by $\varepsilon(1) = 1$, $\varepsilon(\alpha_1 ..., \alpha_r \beta_1 \cdots \beta_s) = 0$.

Property 1 for measuring coalgebras is now verified by observing that the comultiplication given above describes exactly the generalized Leibniz rule for higher order derivations of a graded commutative algebra. The second property is also proved by direct inspection.

One can ask whether the image of p_σ is all of $\bar{T}(B,A)_\sigma$. In the case $A = \mathbb{R}$, this is the statement that Der A acts transitively on $\bar{T}(B, \mathbb{R})_\sigma = \bar{T}A_\sigma$ (see Kostant [6]). The conjecture is also valid for $\bar{T}(A,A)_\sigma$, when σ is an automorphism. See the corollary to the following proposition. In some cases it seems necessary to replace Der $A \oplus$ Der B by the larger Lie superalgebra $A \otimes B \otimes$ (Der $A \oplus$ Der B). See the example of superstrings III.14.

The universal nature of the mapping coalgebra guarantees good behaviour with respect to transformations.

III.11 Proposition

Let A, B, C, D be algebras.

1. There is a multiplication

$$P(A,B) \otimes P(B,C) \rightarrow P(A,C)$$

sending

$$\bar{T}_\sigma(A,B)^k \otimes \bar{T}_\tau(B,C)^h \rightarrow \bar{T}_{\tau\sigma}(A,C)^{k+h} .$$

2. $P(\ ,B)$ is a (contravariant) functor from algebras to coalgebras. That is, for any algebra $A, P(A,B)$ is a coalgebra, and for any algebra homomorphism $f: A \rightarrow C$ there is a coalgebra map $P(f,B): P(C,B) \rightarrow P(A,B)$ such that $P(1,B)$ is the identity, and if $f': C \rightarrow D$ is another algebra map

$$P(f' \circ f, B) = P(f',B) \circ P(f,B).$$

3. Similarly $P(A,*)$ is a (covariant) functor from algebras to coalgebras.

4. If $f: A \rightarrow C$ and $q: B \rightarrow D$ are algebra maps, then the following diagram commutes

$$\begin{array}{ccc} P(C,B) & \xrightarrow{P(f,B)} & P(A,B) \\ {\scriptstyle P(C,g)}\downarrow & & \downarrow{\scriptstyle P(A,q)} \\ P(C,D) & \xrightarrow{P(f,D)} & P(A,D) \end{array}$$

5. The statements in parts 1-3 hold if the morphisms f, f', g etc. are more general elements of the mapping coalgebras $P(A,C)$, $P(C,D)$ and $P(B,D)$ respectively.

6. If A_1, A_2 are algebras, then $P(A_1 \otimes A_2, B) = P(A_1, B) \otimes P(A_2, B)$.

Proofs. (Sketch)

Part 1. The coalgebra structure on $P(A,B) \otimes P(B,C)$ is given as follows. In terms of the comultiplication on $P(A,B)$, $P(B,C)$ respectively,

$$\Delta(\alpha \otimes \beta) = \sum_{(\alpha)(\beta)} (-1)^{|\alpha_{(2)}||\beta_{(1)}|} (\alpha_{(1)} \otimes \beta_{(1)}) \otimes (\alpha_{(2)} \otimes \beta_{(2)}) .$$

Similarly

$$\varepsilon(\alpha \otimes \beta) = \varepsilon_{AB}(\alpha)\, \varepsilon_{BC}(\beta)$$

where $\varepsilon_{AB}, \varepsilon_{BC}$ are the counits for $P(A,B)$, $P(B,C)$.

The measuring maps π_{AB}, π_{BC} enable one to define

$$\rho : P(A,B) \otimes P(B,C) \rightarrow \mathrm{Hom}(A,C)$$

$$\rho(\alpha \otimes \beta) = (-1)^{|\alpha||\beta|} \pi_{BC}(\beta) \pi_{AB}(\alpha) .$$

Check that this map measures:

$$\begin{aligned}<\rho(\alpha\otimes\beta), a_1a_2> &= (-1)^{|\alpha||\beta|}<\pi_{BC}(\beta)\pi_{AB}(\alpha), a_1a_2>\\ &= (-1)^{|\alpha||\beta|}<\pi_{BC}(\beta)<\sum_{(\alpha)}(-1)^{|a_1||\alpha_{(2)}|}<\pi_{AB}(\alpha_{(1)})a_1><\pi_{AB}(\alpha_{(2)}), a_2>\\ &= \sum_{(\alpha)(\beta)}(-1)^{T(\alpha,\beta,a_1)}<\pi_{BC}(\beta_{(1)}), <\pi_{AB}(\alpha_{(1)})\gg<\pi_{BC}(\beta_{(2)}), <\pi_{AB}(\alpha_{(2)}), a_2\gg \qquad \dagger\end{aligned}$$

where

$$\begin{aligned}T(\alpha,\beta,a_1) &= |\alpha||\beta| + |a_1||\alpha_{(2)}| + |<\pi_{AB}(\alpha_{(1)}), a_1>|\ |\beta_{(2)}|\\ &= |\alpha||\beta| + |a_1||\alpha_{(2)}| + |\alpha_{(1)}||\beta_{(2)}| + |a_1||\beta_{(2)}|\ .\end{aligned}$$

This last expression † is

$$\sum(-1)^s<\rho(\alpha_{(1)}\otimes\beta_{(1)}), a_1><\rho(\alpha_{(2)}\otimes\beta_{(2)}, a_2>\ .$$

where s is

$$T(\alpha,\beta,a_1) + |\alpha_{(1)}||\beta_{(1)}| + |\alpha_{(2)}||\beta_{(2)}|$$

Since $|\alpha||\beta| = (|\alpha_{(1)}| + |\alpha_{(2)}|)(|\beta_{(1)}| + |\beta_{(2)}|)$ we get

$$\begin{aligned}s &= T(\alpha,\beta,a_1) + |\alpha_{(1)}||\beta_{(1)}| + |\alpha_{(2)}||\beta_{(2)}|\\ &= |a_1|(\alpha_{(2)} + \beta_{(2)}) + |\alpha_{(2)}||\beta_{(1)}|\ .\end{aligned}$$

Careful examination of the signs will reveal that this is property 1 for measuring maps.

For property 2

$$\begin{aligned}<\rho(\alpha\otimes\beta), 1> &= (-1)^{|\alpha||\beta|}<\pi_{BC}(\beta)\pi_{AB}(\alpha), 1>\\ &= (-1)^{|\alpha||\beta|}<\pi_{BC}(\beta), <\pi_{AB}(\alpha), 1\gg\\ &= (-1)^{|\alpha||\beta|}<\pi_{BC}(\beta), \varepsilon_{AB}(\alpha)\,1>\\ &= (-1)^{|\alpha||\beta|}\varepsilon_{BC}(\beta)\,\varepsilon_{AB}(\alpha)\ .\end{aligned}$$

The sign does not cause problems since the following calculation shows that $\varepsilon_{BC}(\beta) = 0$ if $|\beta| = 1$.

Write $\Delta\beta$ as a sum

$$\Delta\beta = \sum_{\substack{(\beta)\\ |\beta_{(1)}|=0\\ |\beta_{(2)}|=1}}\beta_{(1)}\otimes\beta_{(2)} + \sum_{\substack{(\beta)\\ |\beta_{(2)}|=0\\ |\beta_{(1)}|=f}}\beta_{(1)}\otimes\beta_{(2)} = E + F\ .$$

Since $\beta = \sum_{(\beta)}\varepsilon(\beta_{(1)})\beta_2$ and $|\beta| = 1$,

$$\sum_{|\beta_{(1)}|=1}\varepsilon(\beta_{(1)})\beta_{(2)} = 0$$

$$\beta = \sum_{|\beta_{(1)}| = 0} \varepsilon(\beta_{(1)})\beta_{(2)} \ .$$

But by graded commutativity F equals E with its factors reversed. Thus

$$\begin{aligned}
\varepsilon(\beta) &= \left(\sum_{|\beta_{(1)}| = 0} \varepsilon(\beta_{(1)})\beta_{(2)} \right) \\
&= \sum_{|\beta_{(1)}| = 1} \varepsilon(\beta_{(1)})\beta_{(2)} \\
&= \varepsilon\left(\sum_{|\beta_{(1)}| = 0} \varepsilon(\beta_{(1)})\beta_{(2)} \right) = 0 \, .
\end{aligned}$$

Thus ρ measures, inducing

$$p_B \colon P(A,B) \otimes P(B,C) \to P(A,C) \, .$$

The second part of 1 is immediate from the characterization of elements in $\bar{T}_\sigma(A,B)^k$, $\bar{T}_\tau(B,C)^h$, $\bar{T}_{\tau\sigma}(A,C)^{k+h}$, and the comultiplication in $P(A,B) \otimes P(B,C)$.

The other parts of the proposition are corollaries. The map $P(f,B)$ is $\rho : f \otimes P(C,B) \to P(A,B)$ for example.

Part 4 is a statement that this 'multiplication' is associative. That is, consider the diagram

$$\begin{array}{ccc}
P(A,B) \otimes P(B,C) \otimes P(C,D) & \xrightarrow{1 \otimes P_C} & P(A,B) \otimes P(B,D) \\
\Big\downarrow {\scriptstyle P_B \otimes 1} & & \Big\downarrow {\scriptstyle P_B} \\
P(A,C) \otimes P(C,D) & \xrightarrow{P_C} & P(A,D)
\end{array}$$

If this commutes then 4 will follow. To prove that this commutes observe that the measuring map

$$\rho : P(A,B) \otimes P(B,C) \otimes P(C,D) \to \mathrm{Hom}\,(A,D)$$

$$\rho(\alpha \otimes \beta \otimes \nu) = (-1)^{|\beta||\nu| + |\alpha||\beta| + |\alpha||\nu|} \pi_{CD}(\nu)\pi_{BC}(\beta)\pi_{AB}(\alpha)$$

satisfies

$$\rho = \rho_B(1 \otimes P_C) = \rho_C(P_B \otimes 1) \, .$$

Thus $P_B(1 \otimes P_C)$, $P_C(P_B \otimes 1)$ are two maps into $P(A,D)$ which measure. Thus $P_B(1 \otimes P_C) = P_C(P_B \otimes 1)$ as desired.

Part 6. Write $P_i = P(A_i,B)$. Notice there are natural maps

$$\mathrm{Hom}(A_1,B) \otimes \mathrm{Hom}(A_2,B) \to \mathrm{Hom}(A_1 \otimes A_2, B)$$

$$\mathrm{Hom}(A_1 \otimes A_2, B) \to \mathrm{Hom}(A_1, B)$$

$$\mathrm{Hom}(A_1 \otimes A_2, B) \to \mathrm{Hom}(A_2, B)$$

where the last two are determined by the inclusions

$$A_1 = A_1 \otimes 1 \to A_1 \otimes A_2, \; A_2 = 1 \otimes A_2 \to A_1 \otimes A_2 .$$

Composing the universal measuring maps with these gives maps

$$\tau : P_1 \otimes P_2 \to \mathrm{Hom}(A_1 \otimes A_2, B)$$

$$\rho_i : P(A_1 \otimes A_2, B) \to \mathrm{Hom}(A_i, B) .$$

Check that these maps measure. The universal property then gives coalgebra maps

$$t : P_1 \otimes P_2 \to P(A_1 \otimes A_2, B)$$

$$r_i : P(A_1 \otimes A_2 B) \to P_i .$$

Composing Δ with $r_1 \otimes r_2$ gives

$$s : P(A_1 \otimes A_2) \to P_1 \otimes P_2 .$$

The claim is that s,t are inverses.

Now consider $P(A,A)$. In the light of the previous proposition $P(A,A)$ has an algebra structure. In this way one can discuss the automorphism 'group' of A.

Let $P_{\mathrm{AUT}}(A)$ denote the subcoalgebra of $P(A,A)$ given by

$$P_{\mathrm{AUT}}(A) = \bigoplus_{\sigma \in \mathrm{Aut}(A)} \bar{T}(A,A)_\sigma$$

where $\mathrm{Aut}(A)$ denotes the set of algebras automorphisms.

III.12 Corollary

The subcoalgebra $P_{\mathrm{AUT}}(A)$ is a Hopf algebra and

$$P_{\mathrm{AUT}}(A) = \mathbb{R}(\mathrm{Aut}(A)) \otimes U(\mathrm{Der}(A) .$$

Proof

Notice that $P_{\mathrm{AUT}}(A)$ is a bialgebra (diagram 1 of definition II.6 commutes). This is a statement that multiplication in $P_{\mathrm{AUT}}(A)$ is a coalgebra map: but this follows since multiplication was defined as the universal map associated with measuring coalgebras.

The results now follow from generalizing Sweedler's results on cocommutative pointed bialgebras to the graded commutative setting: 9.2.5 would say that $P_{\mathrm{AUT}}(A)$ is a Hopf algebra, and the exercise in section 13.1 shows that $P_{\mathrm{AUT}}(A) = \mathbb{R}(\mathrm{Aut}(A)) \otimes U(P)$ where P is the set of primitives in $\bar{T}(A,A)_{\mathrm{Id}}$.

It is not hard to verify that P can be identified with $\mathrm{Der}(A)$, using the universal mapping property.

III.13 Proposition

Given an algebra homomorphism $\sigma : A \to B$, define $\mathrm{Diff}(A,B)^0_\sigma$ to be the set of A-module maps

$$\mathrm{Diff}(A,B)^0_\sigma = \{\alpha \in \mathrm{Hom}(A,B) : \alpha(a_1 a_2) = \sigma(a_1)\alpha(a_2)\} .$$

Define $\mathrm{Diff}(A,B)^i_\sigma$ by induction on i

$$\mathrm{Diff}(A,B)^i_\sigma = \{\alpha \in \mathrm{Hom}(A,B) : [\alpha,a] \in \mathrm{Diff}(A,B)^{i-1}_\sigma \text{ for all } a \text{ in } A\} .$$

The map $[\alpha,a]$ here is defined by

$$[\alpha,a]a_1 = \alpha(aa_1) - (-1)^{|\alpha||a|}\sigma(a)\alpha(a_2) .$$

Then

$$\pi(\bar{T}(A,B)^k_\sigma) \subset \mathrm{Diff}(A,B)^k_\sigma .$$

Proof

It can be shown that for p in $\bar{T}(A,B)^k_\sigma$ we can write

$$\Delta p = \sigma \otimes p + \Sigma p_i \otimes q_i$$

with q_i in $\bar{T}(A,B)^{h_i}_\sigma$, $h_i < k$. (Proceed by induction on k.)

Notice that $\bar{T}(A,B)^0_\sigma = \mathbb{R}\sigma \subset \mathrm{Diff}(A,B)^0_\sigma$. Proceed by induction on k to show that for p in $\bar{T}(A,B)^k_\sigma$, $[\pi p,a_1]$ is in $\mathrm{Diff}(A,B)^{k-1}_\sigma$:

$$\begin{aligned}[\pi p,a_1]a_2 &= \pi p(a_1 a_2) - (-1)^{|p||a_1|}\sigma(a_1)\pi p(a_2) \\ &= \Sigma(-1)^{|q_i||a_1|}\pi p_i(a_1)\pi q_i(a_2) .\end{aligned}$$

This calculation shows that $[\pi p,a_1] = \Sigma(-1)^{|q_i||a_1|}\pi p_i(a_1)\pi q_i$ which, since $\mathrm{Diff}(A,B)^{k-1}_\sigma$ is a B module, and since q_i is in $\bar{T}(A,B)^{k-1}_\sigma$, proves the result.

III.14 Examples: Strings, spinning strings and superstrings

One can find examples in almost every type of field theory where one would like a description of the space of smooth maps between manifolds. I will conclude this section by interpreting strings (as presented in Green, Schwarz and Witten [5]) in the context of $P(A,B)$.

Strings. Consider $P(C^\infty(\mathbb{R}^D), C^\infty(\mathbb{R}^2))$. Let x^μ denote the coordinate functions in $C^\infty(\mathbb{R}^D)$. If $s : \mathbb{R}^2 \to \mathbb{R}^D$ is a smooth map, I will, as usual, also write $s : C^\infty(\mathbb{R}^D) \to C^\infty(\mathbb{R}^2)$ for the associated algebra map. Then

$$X^\mu = s(x^\mu) \in C^\infty(\mathbb{R}^2) .$$

If it were desirable to do so, I could also consider $\partial_\alpha \circ s = P_s\partial_\alpha$ in $P(C^\infty(\mathbb{R}^D), C^\infty(\mathbb{R}^2))$. Then

$$\partial_\alpha X^\mu = \partial_\alpha s(x^\mu) = P_s\partial_\alpha(x^\mu) .$$

If X is a Riemann surface, with $X = H/\Gamma$ (H is the upper half plane, Γ is a suitable group acting freely and discontinuously). I can identify $P(C^\infty(\mathbb{R}^D), C^\infty(X))$ as the fixed point set of $P(C^\infty(\mathbb{R}^D), C^\infty(H))$ under the action of Γ.

Spinning strings. A spinning string will be an element p in $P(C^\infty(\mathbb{R}^D), C^\infty(\mathbb{R}^2) \otimes \Lambda\psi)$. Given such a p we get πp in $\mathrm{Hom}(C^\infty(\mathbb{R}^D), C^\infty(\mathbb{R}^2) \otimes \Lambda\psi)$. We can write

$$\pi(p)(x^\mu) = X^\mu + \psi^\mu$$

where $X^\mu(\psi^\mu)$ is the even (odd) part of $\pi(p)(x^\mu)$. (As above x^μ is the coordinate function in $C^\infty(\mathbb{R}^D)$.)

Notice that if πp is an algebra homomorphism (corresponding to a genuine map of graded manifolds) $\psi^\mu = 0$. To get a non-zero ψ^μ I need something in $\bar{T}(C^\infty(\mathbb{R}^D), C^\infty(\mathbb{R}^2) \otimes \Lambda\psi)^1_s$, for example. Explicitly, given $s : C^\infty(\mathbb{R}^D) \to C^\infty(\mathbb{R}^2) \otimes \Lambda\psi$ an algebra map, and $\Sigma f_\mu \frac{\partial}{\partial x}\mu$ in $\mathrm{Der}(C^\infty(\mathbb{R}^D))$ I can build

$$p = s + \theta P_s(\Sigma f_i \frac{\partial}{\partial x_i})$$

with

$$p(x^\mu) = s(x^\mu) + \theta f_\mu .$$

Thus $s(x^\mu) = X^\mu$, and $\theta f_\mu = \psi\mu$. In this setting the property that ψ^μ transforms as a space-time vector and world-sheet spinor becomes a consequence of p being in $\bar{T}(C^\infty(\mathbb{R}^D), C^\infty(\mathbb{R}^2) \otimes \Lambda\psi)^1_s$ rather than an imposed condition. See Green, Schwarz and Witten [5] p.186 (But note, I could choose p in $\bar{T}(C^\infty(\mathbb{R}^D), C^\infty(\mathbb{R}^2) \otimes \Lambda\psi)^k_s$ for $k > 1$.)

Sometimes it is convenient to consider both ψ_+ and ψ_- components of spinning strings. This is achieved by embedding $P(C^\infty(\mathbb{R}^D), C^\infty(\mathbb{R}^2) \otimes \Lambda\psi)$ in $P(C^\infty(\mathbb{R}^D), C^\infty(\mathbb{R}^2) \otimes \Lambda(\psi_+, \psi_-))$.

Super strings. These are now elements p in

$$p(C^\infty(\mathbb{R}^D) \otimes \Lambda(T_1, \ldots, T_s), C^\infty(\mathbb{R}^2))$$

where $s = \frac{D}{2}$. Again, notice that $\pi p(T_i) = 0$ unless p is at least in

$$\bar{T}(C^\infty(\mathbb{R}^D) \otimes \Lambda(T_1, \ldots, T_s), C^\infty(\mathbb{R}^2))^1_\sigma .$$

An example of a map with $\pi p(T_i)$ non-zero is

$$p = \sigma + \underset{\sigma}{P}(\Sigma f_i \frac{\partial}{\partial T_i})$$

for $\sigma : C^\infty(\mathbb{R}^D) \otimes \Lambda(T_1, \ldots, T_s) \to C^\infty(\mathbb{R}^2)$ an algebra homomorphism. Thus

$$\pi(p)(x^\mu) = \sigma(x^\mu) = X^\mu$$

$$\pi(p)\,(T_j) \;=\; \sigma(f_j) \;=\; \theta_j\ .$$

Note that I could also consider 'higher order' super strings,

$$q \;=\; \sigma + P_\sigma(\Sigma f_{ij}\,\frac{\partial}{\partial T_i}\,\frac{\partial}{\partial T_j})\ .$$

IV Application 1. Berezin integration

The aim is to interpret the Berezin integral

$$\oint \theta\, d\theta \;=\; 1$$

$$\oint d\theta \;=\; 0$$

as an integration of a form over a suitable chain. The solution is to use $P(A,B)$ to generalize the notion of chains.

IV.1 Preliminaries on chains

Let Δ^p denote the standard p simplex (the convex subset of $\mathbb{R}^p$ spanned by the standard basis vectors and 0). Let $C_p(X)$ denote the pth chain group; that is, the vector space spanned by maps

$$\sigma : \Delta^p \;\to\; X\,.$$

Since all spaces are smooth, all the maps σ will be assumed to be smooth. The face maps

$$\partial_i : C_p(X) \;\to\; C_{p-1}(X)$$

are given by composing a map σ with the inclusion of Δ^{p-1} as the ith face of Δ^p. The boundary operator $d : C_p(X) \to C_{p-1}(X)$ is $\Sigma(-1)^i\partial_i$. Functoriality is also immediate: if $f : X \to Y$ is a smooth map, one gets a map of chain complexes

$$C_*(f) : C_*(X) \;\to\; C_*(Y)$$

by composing each σ with f.

Preliminaries on Forms

The sheaf Ω of smooth forms on X is usually defined in terms of sections of the exterior bundle of the cotangent bundle. For my purposes it is convenient to use an equivalent algebraic definition.

IV.2 Definition

A differential $\mathbb{Z}\times\mathbb{Z}_2$ graded commutative algebra R is an algebra which has a $\mathbb{Z}\times\mathbb{Z}_2$ grading satisfying

$$ab \;=\; (-1)^{ii' + jj'}ba \;\in\; R_{(i+i',\, j+j')}$$

for a in $R_{(i,j)}$, b in $R_{(i',j')}$. Also R has a differential d of degree (1,0) satisfying

$$d(ab) = (da)b + (-1)^i a(db).$$

The algebra of forms occurs then as a differential $(\mathbf{Z} \times \mathbf{Z}_2)$ graded algebra with a suitable universal property.

IV.3 Definition

Let A be a $\mathbf{Z}(\mathbf{Z}_2$ graded) commutative algebra. Define ΩA to be the $\mathbf{Z}$ $(\mathbf{Z} \times \mathbf{Z}_2)$ graded commutative differential algebra with the universal property that if R is any other $\mathbf{Z}$ $(\mathbf{Z} \times \mathbf{Z}_2)$ graded commutative differential algebra, and $\Psi : A \to R$ is any homomorphism (of $\mathbf{Z}_2$-graded algebras) then there is a unique homomorphism of differential graded algebras making the following diagram commute.

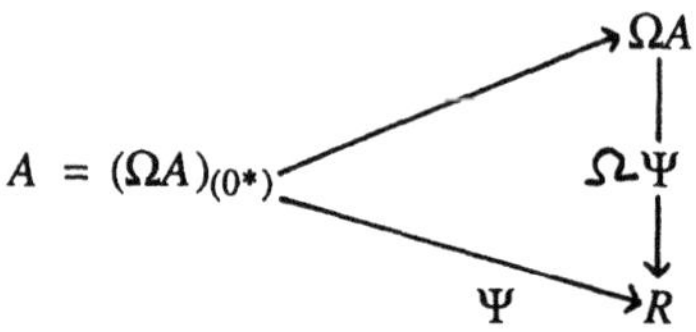

Uniqueness of ΩA is a consequence of the universal property. Existence of ΩA must be demonstrated by constructing ΩA as the free graded commutative algebra generated by A in degree (0,*), dA in (1,*) subject to the relation

$$d(ab) = (da)b + a(db).$$

In the case that $A = C^\infty(X)$, one can verify that $\Omega C^\infty(X) = \Omega(X)$ corresponds to the usual definition by considering $\Omega C^\infty(U)$ for a chart U on X. The universal property of Ω guarantees that forms pull back.

IV.4 Proposition

Let $f : X \to Y$ be a smooth map. Then the associated algebra homomorphism $f : C^\infty(Y) \to C^\infty(X)$ extends to a map of differential graded algebras $\Omega f : \Omega(Y) \to \Omega(X)$.

Proof

This is immediate from the universal property considering $C^\infty(Y) \xrightarrow{f} C^\infty(X) \to \Omega(X)$ as an algebra map into a differential graded algebra.

IV.5 Integration

Since forms pull back, integration of forms on X over chains on X can be defined in terms of integration of forms on the standard simplices (considered as subsets of $\mathbb{R}^n$). Explicitly, define

$$\int_{\Sigma a_i \sigma_i} \omega = \sum a_i \int_{\Delta^p} (\Omega\sigma_i)(\omega)$$

for ω a p-form on X, $\Sigma\, a_i \sigma_i$ in $C_p(X)$, and $\int_{\Delta^p}$ interpreted as the usual integration in $\mathbb{R}^p$.

The change of variable formula can be expressed in the following terms.

IV.6 Proposition

Let $f : X \to Y$ be a smooth map. Then

$$\int_{C_*(f)c} \omega \;=\; \int_c (\Omega f)\omega$$

for ω in $\Omega(Y)$, c in $C_*(X)$.

Proof

Write $c \;=\; \Sigma\, a_i \sigma_i$. Suppose that c is a p-chain, and ω is a p form. Then

$$\begin{aligned}
\int_{C_*(f)c} \omega &= \int_{\Sigma\, a_i\sigma_i \cdot f} \omega \\
&= \Sigma\, a_i \int_{\Delta^p} \Omega(\sigma_i \cdot f)\omega \\
&= \Sigma\, a_i \int_{\Delta^p} \Omega(\sigma_i)(\Omega f)(\omega) \\
&= \int_c (\Omega f)(\omega)\,.
\end{aligned}$$

IV.7 The generalization

One wishes to interpret integration in a broader setting of evaluation. The following diagram is illustrative:

$$\begin{array}{ccccccc}
\left\{\Sigma \text{ second-order derivatives}\right\} & \subset & C^{\infty}(X)^{\circ}_{2} & \cdot & \cdot & \cdot \\[2em]
\left\{\Sigma\xi_i : \xi_i \in TX\right\} & \subset & C^{\infty}(X)^{\circ}_{1} & \cdot & \cdot & \cdot \\[2em]
\left\{\sum_{x_i \in X} a_i x_i\right\} & = & C^{\infty}(X)^{\circ}_{0} = C_0(X) & C_1(X) & C_2(X) & C_3(X) \\[1em]
 & & \int & \int & \int & \int \\[1em]
 & & \Omega^0(X) & \Omega^1(X) & \Omega^2(X) & \Omega^3(X) \\
 & & \| & & & \\
 & & C^{\infty}(X) & & &
\end{array}$$

Here $C^\infty(X)^\circ_i = \underset{x\in X}{\oplus} \bar{T}^i_x$. As I have indicated by writing $\Omega^i(X)$ underneath $C_i(X)$, I know how to integrate $\Omega^i(X)$ over $C_i(X)$. In the special case $i=0$ I know how to 'integrate' $\Omega^\circ(X) = C^\infty(X)$ over the more general 'chains' $C^\infty(X)^\circ_j$. Evidently $\underset{j}{\cup} C^\infty(X)^\circ_j = C^\infty(X)^\circ$. By analogy, I want to replace the dots by 'higher derivatives at p-chains'.

The desired object is clearly the space $P(C^\infty(X), C^\infty(\Delta^p))$. Integration of these 'generalized chains' is a consequence of the following proposition.

IV.8 Proposition

The map $\pi : P(A,B) \to \mathrm{Hom}(A,B)$ extends uniquely to a measuring map

$$\tilde{\pi} : P(A,B) \to \mathrm{Hom}(\Omega A, \Omega B)$$

such that

1) $\tilde{\pi}c$ has $\mathbb{Z}$-degree 0.

2) $[\tilde{\pi}c, d] = 0$.

Proof

The map π gives us an algebra map

$$\pi : A \to \mathrm{Hom}(P(A,B), B).$$

The inclusion $B \to \Omega B$ gives an algebra homomorphism

$$\pi : A \to \mathrm{Hom}(P(A,B), \Omega B).$$

But $\mathrm{Hom}(P(A,B), \Omega B)$ has the structure of a differential graded algebra. Hence there is a unique map of (differential graded) algebras

$$\Omega A \to \mathrm{Hom}(P(A,B), \Omega B).$$

The map $\tilde{\pi}$ is the associated map

$$\tilde{\pi} : P(A,B) \to \mathrm{Hom}(\Omega A, \Omega B).$$

Using $\tilde{\pi}$ we can define integration of a p-form ω on X with respect to any element of $P(C^\infty(X), C^\infty(\Delta^p))$ by setting

$$\int_p \omega = \int_{\Delta^p} \tilde{\pi}(p)(\omega).$$

Moreover, since the constructions used are entirely algebraic there is nothing to stop us using using the above as a definition for integration of forms on a graded manifold.

IV.9 Proposition

1) Let α_i be in Der $C^\infty(\Delta^P)$, β_j in Der $(C^\infty(X))$, and $\sigma : \Delta^p \to X$ be a smooth map. Let $L_{\alpha_i}(L_{\beta_j})$ denote the Lie derivative with respect to α_i, (β_j). Then if ω is a form on X,

$$\tilde{\pi}(P_\sigma \alpha_1 \cdots \alpha_r \beta_1 \cdots \beta_s)\omega$$

$$= L_{\alpha_1} \cdots L_{\alpha_r} \Omega\sigma L_{\beta_1} \cdots L_{\beta_s} \omega .$$

Recall that P_σ is the map $P_\sigma : U(A,B) \to \overline{T}(A,B)_\sigma$.

2) Let $A = C^\infty(\mathbb{R}^r) \otimes \Lambda(\theta_1, ..., \theta_s)$. Then

a) $$\int_{P_u\left(\frac{\partial}{\partial\theta_s} \cdots \frac{\partial}{\partial\theta_1}\right)} \theta_1 \cdots \theta_s = 1$$

where u is any point in $\mathbb{R}^r$.

b) $$\int_{P_{\tilde{\sigma}}\left(\frac{\partial}{\partial\theta_s} \cdots \frac{\partial}{\partial\theta_1}\Big|_0\right)} \theta_1 \cdots \theta_s\, \omega = \int_\sigma \omega$$

for $\sigma : \Delta^P \to X$ a smooth map, $\tilde{\sigma} : C^\infty(X) \otimes \Lambda(\theta_1 \cdots \theta_s) \to C^\infty(\Delta^p)$ the associated algebra homomorphism sending all the θ's to 0, and ω a p-form on X.

c) $$\int_{P_{\tilde{\sigma}}\left(\frac{\partial}{\partial\theta_s} \cdots \frac{\partial}{\partial\theta_1}\right)} \theta_1 \cdots \theta_s\, d\theta_1^{i_1} \cdots d\theta_s^{i_s}\, \omega$$

$$= 0 \quad \text{unless } i_h = 0 \text{ for all } h.$$

d) $$\int_{P_{\tilde{\sigma}}\left(\frac{\partial^{j_s}}{\partial\theta_s} \cdots \frac{\partial^{j_1}}{\partial\theta_1}\right)} \theta_1^{k_1} \cdots \theta_s^{k_s}\, \omega$$

$$= \delta_{j_1}^{k_1} \cdots \delta_{j_s}^{k_s} \int_\sigma \omega.$$

3) If $f : B(Y) \to A(X)$ is a map of graded manifolds, and c is in $P(A, C^\infty(\Delta^q))$, ω is a q form on $B(Y)$, then

$$\int_{P(f,C^\infty(\Delta^q))c} \omega = \int_c \tilde{\pi}(P(f, C^\infty(\Delta^q))) .$$

4) If f is an automorphism of $C^\infty(\Delta^r) \otimes \Lambda(\theta_1 \cdots \theta_s)$, the change of variables formula (3) can be written in terms of the Berezinian

$$\int_c \theta_1 \cdots \theta_s\, dx_1 \cdots dx_r$$

$$= \int_{P(f^{-1},C^\infty(\Delta^r))c} \tilde{\pi}(P(f, C^\infty(\Delta^r)))\,(\theta_1 \cdots \theta_s\, dx_1 \cdots dx_r^1)$$

where $c = P_{\mathrm{Id}} \frac{\theta}{\partial\theta_s} \cdots \frac{\partial}{\partial\theta_1}$.

Also if $\theta'_t = f(\theta_i), x'_i = f(x_i)$ and $J = \begin{pmatrix} \frac{\partial x'_i}{\partial x_j} & \frac{\partial x'_i}{\partial\theta_j} \\ \frac{\partial\theta'_i}{\partial x_j} & \frac{\partial\theta'_i}{\partial\theta_j} \end{pmatrix} = \begin{bmatrix} A & B \\ C & D \end{bmatrix}$

$$\begin{aligned}\tilde{\pi}(P(f, C^\infty(\Delta^r)))\, \theta_1 \cdots \theta_s\, dx_1 \cdots dx_1 &= \theta'_1 \cdots \theta'_s\, dx'_1 \cdots dx'_r \\ &= \det A \det D\, \theta_1 \cdots \theta_s\, dx_1 \cdots dx_r \\ &\quad + \text{ terms involving } d\theta_j .\end{aligned}$$

If $J^{-1} = \begin{pmatrix} P & Q \\ R & S \end{pmatrix}$, then $S = (D - CA^{-1}B)^{-1}$ and

$$\begin{aligned}P(f^{-1}, C^\infty(\Delta^r))c &= P_{\mathrm{Id}} \frac{\partial}{\partial\theta_1} \cdots \frac{\partial}{\partial\theta_s} \bullet f^{-1} \\ &= P_{\mathrm{Id}} f^{-1} \frac{\partial'}{\partial\theta_1} \cdots \frac{\partial'}{\partial\theta_s} \\ &= P_{\mathrm{Id}} f^{-1} \left\{ \mathrm{Det}\,(D - CA^{-1}B)^{-1} \frac{\partial}{\partial\theta_s} \cdots \frac{\partial}{\partial\theta_1} \right. \\ &\qquad \left. + \text{ terms missing at least one } \frac{\partial}{\partial\theta_i} \text{ factor} \right\}.\end{aligned}$$

Proof

Parts 1-3 are exercises in notation. Part 4 requires a little linear algebra. Write $U = \begin{bmatrix} AO \\ OD \end{bmatrix}$, $N = \begin{bmatrix} OB \\ CO \end{bmatrix}$. Thus $J = U + N$, where N is nilpotent, and

$$J^{-1} = \left(\sum_n (-U^{-1}N)^n \right) U^{-1} .$$

The bottom right hand corner of this is $\left(\sum_n (D^{-1}CA^{-1}B)^n \right) D^{-1}$. Using the series expression for inverses again shows that

$$\begin{aligned}S &= (1 - D^{-1}CA^{-1}B)^{-1} D^{-1} \\ &= (D^{-1}(D - CA^{-1}B))^{-1} D^{-1} \\ &= (D - CA^{-1}B)^{-1}\end{aligned}$$

as desired.

V Application 2. The automorphism group of the graded Riemann sphere

As a concrete example one can calculate the 'automorphism group' of the graded Riemann sphere. Having done so, one can look at the 'subgroup' which preserves the complex structure. The groups and algebras which arise in doing so may explain why graded Riemann surfaces (or super-Riemann surfaces) should be useful in modelling physics. Most of this material is covered in more detail in [3].

VI The graded Riemann sphere

The Riemann sphere $S = \mathbb{C} \cup \{\infty\}$ can also be interpreted as complex projective space. Explicitly, identify the line through the point (a,b) in $\mathbb{C}^2$ with the point a/b in $\mathbb{C} \cup \{\infty\}$. As such, S has a tautological complex line bundle over it, with the fibre at a point z being the complex line corresponding to z. Call this bundle E.

Now construct ΛE. As a bundle it is isomorphic to $E \oplus \mathbb{C}$, where $\mathbb{C}$ stands for the trivial complex line bundle. One can consider analytic, meromorphic, or smooth sections of ΛE, denoted by Γ_ω, Γ_m and Γ respectively. The grading on ΛE shows a grading on Γ (and Γ_ω, Γ_m) with $\Gamma_0(\Gamma_{\omega 0}, \Gamma_{m0})$ being the smooth (holomorphic, meromorphic) complex valued functions, and $\Gamma_1(\Gamma_{\omega 1}, \Gamma_{m1})$ the smooth (holomorphic, meromorphic) sections of E.

The graded Riemann sphere is the pair (S, Γ). This is an ordinary graded manifold in the sense of Kostant except that the ground field is the complex (rather than the real) numbers. The sheaf Γ_ω is a subsheaf of Γ and Γ_m.

The bundle E is a spin bundle. We can choose an explicit isomorphism of $\Gamma_{\omega 1} \otimes \Gamma_{\omega 1}$ (tensored over the holomorphic functions) with the sheaf of holomorphic 1 forms, Ω^1.
Thus $\Omega^1 = \{f dz,\ f \text{ holomorphic}\}$.

For example, define

$$k : \Gamma_{\omega 1} \otimes \Gamma_{\omega 1} \to \Omega^1$$

by setting

$$k : \Gamma_{\omega 1}(S - \{\infty\}) \otimes \Gamma_{\omega 1}(S - \{\infty\}) \to \Omega^1(S - \{\infty\})$$

$$k(\theta \otimes \theta) = dz$$

where θ is the section of $S - \{\infty\} = \mathbb{C}$ given by $\theta(z) = (z, 1)$. It can be verified that this defines a map of sheaves.

V.2 Maps of ringed spaces

The set of graded manifold maps is simply the set of algebra maps $\sigma : \Gamma(S) \to \Gamma(S)$. Any graded manifold map gives rise to a map of ringed spaces: that is, a map of smooth manifolds

$$\sigma : S \to S$$

and algebra homomorphisms

$$\sigma_U : \Gamma(\sigma^{-1}(U)) \to \Gamma(U)$$

which commute appropriately with restrictions.

Now I can state my questions: What are the automorphisms of $\Gamma(S)$ which, considered as maps of ringed spaces, preserve Γ_ω?

V.3 Proposition

1. There is a map $K : SL(2,\mathbb{C}) \to \mathrm{Aut}(\Gamma(S))$ determined by

$$\left(K\begin{bmatrix} a & b \\ c & d \end{bmatrix} f\right)(z) = f\left(\frac{az+b}{cz+d}\right)$$

$$\left(K\begin{bmatrix} a & b \\ c & d \end{bmatrix}(\theta)\right)(z) = \frac{1}{cz+d}\,\theta(z)$$

over $S - \{\infty, \frac{-d}{c}\}$.

2. If σ is an automorphism of $\Gamma(S)$ which, considered as a map of ringed spaces makes the following diagrams commute:

$$\begin{array}{ccc} \Gamma(\sigma^{-1}(U)) & \xrightarrow{\sigma_U} & \Gamma(U) \\ \uparrow & & \uparrow \\ \Gamma_\omega(\sigma^{-1}(U)) & \longrightarrow & \Gamma_\omega(U) \end{array}$$

$$\begin{array}{ccc} \Gamma_{\omega 1}(\sigma^{-1}(U) \otimes \Gamma_{\omega 1}(\sigma^{-1}(U)) & \longrightarrow & \Omega^1(\sigma^{-1}(U)) \\ \downarrow{\scriptstyle \sigma_u \otimes \sigma_u} & & \downarrow{\scriptstyle D\sigma} \\ \Gamma_{\omega 1}(U) \otimes \Gamma_{\omega 1}(U) & \longrightarrow & \Omega^1(U) \end{array}$$

then σ is in the image of K.

3. $\Gamma_m(S) = R \oplus R\theta$, where R is the algebra of rational functions. $\mathrm{Aut}(\Gamma_m(S))$ is the semi direct product of the group of Möbius transformations ($= SL(2,\mathbb{C})/_{\{I,-I\}}$ = the Lorentz group) with the group R^* of non-zero rational functions.

4. Let Ω_m^1 denote the meromorphic 1-forms on S. An automorphism σ of $\Gamma_m(S)$ which satisfies

$$\begin{array}{ccc} \Gamma_m(S) \otimes \Gamma_m(S) & \xrightarrow{k} & \Omega^1_m(S) \\ \sigma \otimes \sigma \downarrow & & \downarrow d\sigma \\ \Gamma_m(S) \otimes \Gamma_m(S) & \xrightarrow{k} & \Omega^1_m(S) \end{array}$$

is in the image of K.

With the exception of part 3, which is not difficult, the proofs are in [3].

If one accepts this proposition as motivation for considering $R \otimes \Lambda\theta$ as a suitable algebra of superfields one can also consider the derivations of $R \otimes \Lambda\theta$ which preserve k. I can get the following result and its corollary.

V.4 Proposition

1. Let L_0 denote the Lie subalgebra of (even) derivations γ in $(\mathrm{Der}(R \otimes \Lambda\theta))_0$ which preserve k: that is, all γ for which the following diagram commutes.

$$\begin{array}{ccc} R\theta \otimes R\theta & \xrightarrow{k} & Rdz \\ \gamma \otimes 1 + 1 \otimes \gamma \downarrow & & \downarrow d\gamma \\ R\theta \otimes R\theta & \xrightarrow{k} & Rdz \end{array}$$

Then

$$L_0 = \left\{ p \frac{\partial}{\partial z} + \frac{1}{2} \left(\frac{\partial p}{\partial z} \right) \theta \frac{\partial}{\partial \theta} : p \text{ in } R \right\}.$$

2. Let r be a complex number and set

$$L_1^r = \left\{ p \frac{\partial}{\partial \theta} - rp \frac{\partial}{\partial z} : p \text{ in } R \right\}.$$

Then $L_0 \oplus L_1^r$ form a Lie superalgebra. Moreover if W is a subspace of $\mathrm{Der}(R \otimes \Lambda\theta)_1$ such that $L_0 \oplus W$ is a Lie superalgebra, then $W \subset L_1^r$ for some r.

Proof

1. Writing $\gamma(\theta) = h\theta$ and chasing the diagram shows that γ is in L_0 if and only if $\frac{\partial}{\partial z}(\gamma(z)) = 2h$.

2. A general element of $\mathrm{Der}(R \otimes \Lambda\theta)_1$ is of the form $\tau = p \frac{\partial}{\partial \theta} + q\theta \frac{\partial}{\partial z}$. Computing $[\tau, \tau]$ explicitly shows that $[\tau, \tau]$ is in L_0 if and only if $pq' = p'q$, or $\left(\frac{p}{q} \right)$ is a constant $-r$.

Computing $[\tau,\sigma]$ for $\tau = p\frac{\partial}{\partial\theta} - rp\theta\frac{\partial}{\partial z}$, $\sigma = q\frac{\partial}{\partial\theta} - sq\theta\frac{\partial}{\partial z}$ shows that $[\tau,\sigma]$ is in L_0 if and only if

$$(s-r)pq' = (s-r)p'q \ .$$

Thus either $s = r$ or $\left(\frac{p}{q}\right)$ is constant. But if $\left(\frac{p}{q}\right)$ is constant,

$$\left[z\frac{\partial}{\partial z} + \tfrac{1}{2}\theta\frac{\partial}{\partial\theta}, \tau\right] = (z^2p' - zp)\frac{\partial}{\partial\theta} + r(z^2p' - zp)\,\theta\frac{\partial}{\partial z}$$

is also in W and $(z^2p' - zp)$ and p are not both constant multiples of q.

This shows that $s = r$, and that any subspace W which together with L_0 forms a Lie superalgebra must be contained in some L_1^r.

V.5 Corollary

1. The charge-0 Virasoro algebra has a representation in $\mathrm{Der}(R \otimes \Lambda\theta)_0$

$$L_m = -z^{m+1}\frac{\partial}{\partial z} - \frac{(m+1)}{2} z^m\,\theta\frac{\partial}{\partial\theta}.$$

2. The Neveu-Schwarz Super Virasoro algebra has a representation

$$G_\rho = z^{\rho+1/2}\frac{\partial}{\partial\theta} - z^{\rho+1/2}\,\theta\frac{\partial}{\partial z}$$

for ρ half integral.

While you may be disappointed that only the charge-0 Virasoro algebras have a representation in $\mathrm{Der}(R \otimes \Lambda\theta)$, it is in keeping with the spirit of quantization that the unquantized Lie algebra should have charge 0, and should be represented as derivations.

The Ramond sector can be identified with the derivations of $\hat{R} \otimes \Lambda\theta$, where $\hat{R}$ is the algebra of rational functions and their square roots.

On balance I would like to recommend the Hopf algebra

$$\mathbb{R}(SL(2,\mathbb{C})) \otimes U(L_0 + L_1^1) \ .$$

This is the automorphism group, in the sense of P_{AUT} of the graded Riemann sphere.

VI Functions on the space of maps

Without attempting a systematic study of the 'smooth' functions on the space of maps, it is worthwhile giving some examples of types of functions which ought to be smooth. These illustrate the richness of this algebra by comparison with the algebra of functions on an ordinary manifold. I will begin by describing the simplest types of functions of the space of smooth maps between manifolds X and Y. This will motivate more general types of functions.

VI.1 The basic case

Suppose X,Y are smooth manifolds. Any pair (x,f), x in X, f in $C^\infty(Y)$ give rise to a smooth map on $M(X,Y)$ as follows. For σ in $M(X,Y)$ define $<\sigma,(x,f)> = f(\sigma(x))$:

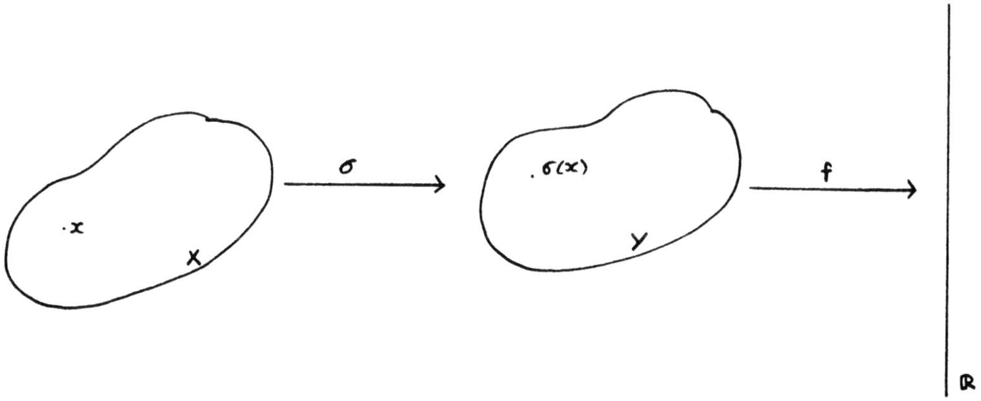

Remark. By inspection $(x,f)(x,g) = (x,fg)$ and $(x,1)=(y,1)=1$, $(x,f)+(x,g)=(x,f+g)$, $(x,a)+(y,b)=(x,a+b)=(y,a+b)=a+b$. This set of maps however is not closed under addition or multiplication. The following generalization provides a space which is closed under addition.

VI.2 Generalization from X to $C^\infty(X)^\circ$

For α in $C^\infty(X)^\circ$, f in $C^\infty(Y)$ and $\sigma : X \rightarrow Y$ define

$$<\sigma, (f,\alpha)> \;=\; <\sigma f, \alpha>$$

considering σ as an algebra homomorphism $\sigma : C^\infty(Y) \rightarrow C^\infty(X)$. More generally elements of $C^\infty(Y) \otimes C^\infty(X)^\circ$ evaluate on elements of $P(C^\infty(Y), C^\infty(X))$ via

$$<\sigma, \Sigma f_i \otimes \alpha_i> \;=\; \sum <\pi(\sigma)f_i, \alpha_i>$$

$$= \sum <\sigma f_i, \alpha_i>.$$

To simplify notation I writre pf_i for $\pi(p)f_i$ from now on.

So far I have a linear map

$$E_1 : C^\infty(Y) \otimes C^\infty(X)^\circ \rightarrow P(C^\infty(Y), C^\infty(X))'$$

which I can represent schematically

$$\begin{array}{ccccc} C^\infty(Y) & \to & C^\infty(X) & \to & \mathbb{R} \\ f & \to & \sigma f & \to & <\sigma f, \alpha> \end{array}$$

VI.3 Generalization from $C^\infty(Y)$ to $S(C^\infty(Y) \otimes P(C^\infty(X), C^\infty(X))$

Here S stands for the symmetric algebra, in this instance the symmetric algebra on the linear space $C^\infty(Y) \otimes P(C^\infty(X), C^\infty(X))$. Write this as $S(C^\infty(Y) \otimes P(X))$. We have a linear map

$$K : C^\infty(Y) \otimes P(X) \to \mathrm{Hom}(P(C^\infty(Y), C^\infty(X)), C^\infty(X))$$

taking an element $f \otimes p$ in $C^\infty(Y) \otimes P(X)$ to $p\sigma f$. Since $\mathrm{Hom}(P(C^\infty(Y), C^\infty(X))$ is a commutative algebra, there is a unique algebra homomorphism

$$\tilde{K} : S(C^\infty(Y) \otimes P(X)) \to \mathrm{Hom}(P(C^\infty(Y), C^\infty(X)), C^\infty(X))$$

extending K above. Use $\tilde{K}$ to define

$$E_2 : S(C^\infty(Y) \otimes P(X)) \otimes C^\infty(X)^\circ \to P(C^\infty(Y), C^\infty(X))' .$$

Schematically for f in $C^\infty(Y)$, p in $P(X)$, α in $C^\infty(X)^\circ$, and σ in $P(C^\infty(Y), C^\infty(X))$, E_2 is defined by

$$\begin{array}{ccccccc} C^\infty(Y) & \to & C^\infty(X) & \to & C^\infty(X) & \to & \mathbb{R} \\ f & \to & \sigma f & \to & p\sigma f & \to & <p\sigma f, \alpha> \end{array}$$

In detail, if $\prod_i (f_i \otimes p_i)$ is in $S(C^\infty(Y) \otimes P(X))$, E_2 is defined by

$$<\sigma, \prod_i^T (f_i \otimes p_i) \otimes \alpha> = \sum_{(\sigma)} <(p_1\sigma_{(1)}f_1)(p_2\sigma_{(2)}f_2) \cdots (p_T\sigma_{(T)}f_T), \alpha> .$$

Still, $S(C^\infty(Y) \otimes P(X)) \otimes C^\infty(X)^\circ$ fails to be an algebra.

VI.4 Generalization to $S(S(C^\infty(Y) \otimes P(X)) \otimes C^\infty(X)^\circ)$

The latest version, E_2, describes a linear map from a vector space into the commutative algebra $P(C^\infty(Y), C^\infty(X))'$. Using the universal property of the symmetric algebra again, I get an algebra homomorphism

$$E_3 : S(S(C^\infty(Y) \otimes P(X)) \otimes C^\infty(X)^\circ) \to P(C^\infty(Y), C^\infty(X))' .$$

The image of E describes a perfectly good subalgebra of functions of $P(C^\infty(Y), C^\infty(X))$ which ought to be smooth. If X is a point this reduces to the image of $S(S(C^\infty(Y)))$ in $(C^\infty(Y)^\circ)'$, which is readily seen to be just $C^\infty(Y)$.

This algebra of functions is not big enough in that I cannot yet describe even simple functions on *M(X,Y)* commonly in use in physics. I need to allow integration as a generalization of evaluation of

$C^\infty(X)$ on $C^\infty(X)^\circ$, as in Application 1.

VI.5 Generalization from $C^\infty(Y) \otimes C^\infty(X)^\circ$ to $\Omega^q(Y) \otimes \Omega^p(X) \otimes P(C^\infty(X), C^\infty(\Delta^{p+q}))$

In Application 1 I have indicated how the integration of forms over chains generalized the evaluation of functions of points. By analogy with E, then, we get a map

$$E_4 : \Omega^q(Y) \otimes \Omega^p(X) \otimes P(C^\infty(X), C^\infty(\Delta^{p+q})) \rightarrow P(C^\infty(Y), C^\infty(X))'$$

given by

$$<\sigma, \omega \otimes \nu \otimes a> \; = \; \int_a (\tilde{\pi}\sigma(\omega)) \wedge \nu$$

for σ in $P(C^\infty(Y), C^\infty(X))$, ω in $\Omega^q(Y)$, ν in $\Omega^p(X)$, a in $P(C^\infty(X), C^\infty(\Delta^{p+q}))$. Write $\sigma(\omega)$ for $\tilde{\pi}\sigma(\omega)$ from now on.

Schematically

$$\begin{array}{ccccccc} \Omega(Y) & \rightarrow & \Omega(X) & \rightarrow & \Omega(X) & \rightarrow & \mathbb{R} \\ \omega & \rightarrow & \sigma\omega & \rightarrow & \sigma\omega \wedge \nu & \rightarrow & \int_a \sigma\omega \wedge \nu \end{array}.$$

Just as E_1 generalizes to E_2 the map

$$K : \Omega(Y) \otimes P(X) \rightarrow \mathrm{Hom}(P(C^\infty(Y), C^\infty(X)), \Omega X)$$

$$<\sigma, \omega \otimes p> \; = \; \sigma p(\omega)$$

extends to an algebra homomorphism

$$\tilde{K} : \hat{S}(\Omega(Y) \otimes P(X)) \rightarrow \mathrm{Hom}(P(C^\infty(Y), C^\infty(X)), \Omega X)$$

where $\hat{S}$ is the graded symmetric algebra on the graded vector space $\Omega(Y) \otimes P(X)$, i.e.

$$\hat{S}(\Omega(Y) \otimes P(X)) \; = \; S(\Omega^{\mathrm{even}}(Y) \otimes P(X)) \otimes \Lambda(\Omega^{\mathrm{odd}}(Y) \otimes P(X)).$$

In turn, use $\tilde{K}$ to define

$$E_5 : \bigoplus_k \hat{S}(\Omega(Y) \otimes P(X)) \otimes \Omega(X) \otimes P(C^\infty(X) \otimes C^\infty(\Delta^k)) \rightarrow P(C^\infty(Y), C^\infty(X))'.$$

Again this fails to be an algebra. As E_3 generalizes E_2, so we can extend E_5 to an algebra map

$$E_6 : \hat{S}(\bigoplus_k \hat{S}(\Omega Y) \otimes P(X)) \otimes \Omega(X) \otimes P(C^\infty(X) \otimes C^\infty(\Delta^k))) \rightarrow P(C^\infty(Y), C^\infty(X))'.$$

The only remaining generalization I wish to propose is that of considering graded manifolds. As the constructions are identical I will not repeat the steps. Observe that if X is a point, again the image of E_6 is just $C^\infty(Y)$.

VI.6 Example 1 The bosonic string action.

Consider 2.1.11 of Green, Schwarz and Witten [5]

$$S = -\frac{T}{2}\int d^2\sigma\sqrt{h}\, h^{\alpha\beta}(\sigma)g_{\mu\nu}(X)\partial_\alpha X^\mu\,\partial_\beta X^\nu\,.$$

I want to identify this action as a function on $P(C^\infty(\mathbb{R}^P), C^\infty(M))$ (here M is the world sheet) of the type described above (the image of E_6). I can identify each of the objects in the integrand as follows:

$d^2\sigma\sqrt{h}$	is a chosen top form on M.
$h^{\alpha\beta}$	is in $C^\infty(M)$ for each α,β
$g_{\mu\nu}$	is in $C^\infty(\mathbb{R}^D)$ for each μ,ν
$X^\mu = s(x^\mu)$	is the image of the coordinate function x^μ on $\mathbb{R}^D$ under the string map $s:\mathbb{R}^2 \rightarrow \mathbb{R}^D$ in $P(C^\infty(\mathbb{R}^D), C^\infty(M))$
$\partial_\alpha,\partial_\beta$	are derivations of $C^\infty(\mathbb{R}^2)$ hence elements of $P(M)$

Implicitly the integration is over a chain c in $P(C^\infty(M), C^\infty(\Delta^2))$ representing the top homology class of M.

Since $g_{\mu\nu}\otimes 1, X^\mu\otimes\partial_\alpha, X^\nu\otimes\partial_\beta$ are all in $\Omega(\mathbb{R}^D)\otimes P(M)$

$$(g_{\mu\nu}\otimes 1)\,(X^\mu\otimes\partial_\alpha)\,(X^\nu\otimes\partial_\beta) \;=\; g_{\mu\nu}\,\partial_\alpha X^\mu\,\partial_\beta X^\nu$$

is in $\hat{S}(\Omega(Y)\otimes P(X))$. Also $d^2\sigma\sqrt{h}\,h_{\alpha\beta}$ is in $\Omega^2(M)$ for each α,β. Thus

$$S \;=\; E_5(-\frac{T}{2}(g_{\mu\nu}\otimes 1)\,(X^\nu\otimes\partial_\beta)\;\otimes\;(d^2\sigma\sqrt{h}\,h_{\alpha\beta})\;\otimes\;c)\,.$$

VI.7 Example 2. The spinning string action.

Here the world sheet is a (2,2) real graded manifold *(M,A)*. Spinning strings are maps from the super-world sheet into $\mathbb{R}^D$, that is, elements s in $P(C^\infty(\mathbb{R}^D), A)$. As before I assume I have coordinates on *(M,A)*, ξ_0,ξ_1 even coordinate functions and Ψ_+,Ψ_- odd coordinate functions. With respect to these coordinates I have differential operators $\partial_0 = \frac{\partial}{\partial\xi_0}$, $\partial_1 = \frac{\partial}{\partial\xi_1}$ and $\frac{\partial}{\partial\Psi_+} = \partial_+$, $\frac{\partial}{\partial\Psi_-} = \partial_-$.

The two dimensional Dirac matrices ρ^0, ρ^2 (4.1.3 in Green, Schwartz and Witten [5]) can be represented as differential operators

$$\rho^\alpha = \Sigma \rho^\alpha_{ab} \Psi_a \partial_b .$$

Here *a,b* run over +, - . The Dirac operator itself becomes

$$\not\partial = -i\rho^\alpha \partial_\alpha = -i\left(\Sigma \rho^\alpha_{ab} \Psi_a \partial_b \partial_\alpha\right).$$

Here α runs over 0,1. Notice that $\not\partial$is in *P(A,A)*; thus a pseudo-differential operator becomes a graded differential operator.

I interpret the object $(\bar{\chi}\Psi)$ to behave like a real valued function on *M*. By direct computation it appears that

$$\bar{\chi}\Psi = i\,\partial_+\partial_-\chi\Psi .$$

I can now identify the action S_2 (4.3.78) as a function in the image of E_5.

$$S_2 = -\frac{1}{2\pi}\int d^2\sigma\, e\left\{h^{\alpha\beta}\partial_\alpha X^\mu \partial_\beta X_m - i\,\overline{\Psi^\mu}\rho^\alpha\partial_\alpha\Psi_\mu\right\}.$$

The first part of the integrand $d^2\sigma e h^{\alpha\beta}\partial_\alpha X^\mu\partial_\beta X_\mu$ is as in example 1, with the simplification that $g_{\mu\nu}$ is now the constant Lorentz metric on $\mathbb{R}^D$. For the second part, observe that Ψ^μ is the odd part of $s(x^\mu)$, hence

$$-\frac{1}{2\pi}\int d^2\sigma\, e\left\{-i\,\overline{\Psi}^\mu\,\rho^\alpha\,\partial_\alpha\,\Psi_\mu\right\}$$

$$= E_5\left(-\frac{1}{2\pi}(x^\mu\otimes 1)(\chi_\mu\otimes\not\partial)\otimes(d^2\sigma e)\otimes(i\,\partial_+\partial_-c))\right).$$

For the forms S_3 and S_4, new fields χ_α, $\alpha = 0,1$, are introduced, each of which may be interpreted as an odd element of *A*. Thus

$$S_3 = -\frac{1}{\pi}\int d^2\sigma\, e(\bar{\chi}_\alpha\,\rho^\beta\,\rho^\alpha\,\Psi^\mu\,\partial_\beta\,X_\mu)$$

$$= E_5\left(\frac{1}{\pi}(x^\mu\otimes\rho^\alpha\rho^\beta)(x_\mu\otimes\partial_\beta)\otimes(\chi_\alpha\, d^2\sigma e)\otimes(i\,\partial_+\partial c)\right).$$

For S_4 observe that $\bar{\chi}_\alpha\rho^\beta\rho_\alpha\chi_\beta$ is an even field in *A*. Hence

$$S_4 = \frac{1}{4\pi}\int d^2\sigma\, e\,\overline{\Psi}_\mu\,\Psi^\mu\,\bar{\chi}_\alpha\,\rho^\beta\,\rho^\alpha\,\chi_\beta$$

$$= E_5\left(\frac{1}{4\pi}(x_\mu \otimes 1)\,(x^\mu \otimes 1) \otimes (\bar{\chi}_\alpha\, \rho^\beta\, \rho^\alpha\, \chi_\beta\, d^2\, \sigma\, e) \otimes (i\, \partial_+\, \partial_-\, c)) \right).$$

VI.8 Example 3. Super strings.

Here the world sheet is an ordinary manifold M and spacetime is super, $(\mathbb{R}^D, C^\infty(\,) \otimes \Lambda(T_1, ..., T_s))$ where $s = 2^{D/2}$. Strings are elements s in $P(C^\infty(\mathbb{R}^D) \otimes \Lambda(T), C^\infty(M))$. As before $X^\mu = s(x^\mu)$ where x^μ is the coordinate projection in $C^\infty(\mathbb{R}^D)$. In addition

$$\theta_i = s(T_i),$$

thus θ_i is a function on M. The expression $\bar{\theta}\Gamma^\mu\partial_\alpha\theta$ appears to be shorthand notation

$$\bar{\theta}\Gamma^\mu\partial_\alpha\theta = \sum_{i,j} \bar{\theta}_i(\Gamma^0\Gamma^\mu)_{ij}\partial_\alpha\theta_j = \Sigma(\Gamma^0\Gamma^\mu)_{ij}\bar{\theta}_i\partial_\alpha\theta_j .$$

(Perhaps I ought to allow complex-valued functions to make sense of $\bar{\theta}_i$.) Thus interpreting 5.1.20 and 5.1.22 in the case $N = 1$

$$S_1 + S_2 = -\frac{1}{2\pi}\int d^2\sigma\sqrt{h}\, h^{\alpha\beta}\Pi_\alpha\, \Pi_\beta + \frac{1}{\pi}\int d^2\sigma \left\{ -i\, \varepsilon^{\alpha\beta}\partial_\alpha X^\mu\bar{\theta}\Gamma_\mu\partial_\beta\theta \right\}$$

$$= E_5\Bigg(-\frac{1}{2\pi}\Big\{ (x^\mu \otimes \partial_\alpha + (\Gamma^0\Gamma^\mu)_{ij}\,(T_i \otimes 1)\,(T_j \otimes \partial_\alpha))$$

$$\bullet\,(x_\mu \otimes \partial_\beta + (\Gamma^0\Gamma_\mu)_{hk}\,(T_h \otimes 1)\,(T_k \otimes \partial_\beta))$$

$$\otimes\ (d^2\sigma(\sqrt{h}\, h^{\alpha\beta} + 2i\, \varepsilon^{\alpha\beta})) \Big\} \Bigg).$$

The object $\varepsilon^{\alpha\beta}$ is understood to be an element of $C^\infty(M)$ for each α, β in $\{0,1\}$.

I have demonstrated that some functions on spaces of maps which are of interest to physicists can be interpreted in this way. While the notation is unacceptably cumbersome, I hope that the conceptual framework may prove useful.

References

1. Abraham, R. and Marsden, J. *Foundations of Mechanics*. 2nd Ed. Benjamin Cummings, New York, 1978.

2. Batchelor, M. 'Graded manifolds and vector bundles: a functorial correspondence'. *J. Math. Phys.* Vol.26, 1578-1582. 1985.

3. Batchelor, M. and Bryant, P. 'Graded Riemann surfaces'. Commun. Math. Phys. 114, 243-255. 1988.

4. Cohn, P.M. *Algebra*. Vol.2. Wiley, London, 1974.

5. Green, M., Schwarz, J. and Witten, E. *Superstring Theory*. Vol.1. Cambridge University Press, 1987.

6. Kostant, B. 'Graded manifolds, graded Lie groups and pre-quantization', in *Differential Geometric Methods in Mathematical Physics*. Springer Lecture Notes in Maths, Vol.570, 1977.

7. Sweedler, M. *Hopf Algebras*. Benjamin, New York, 1969.

PATH INTEGRATION IN SUPERSPACE

Alice Rogers
Department of Mathematics, King's College
Strand, London WC2R 2LS

In these lectures a rigorous analytic theory of path integration in superspace is described, together with applications to various systems in fermionic and supersymmetric quantum mechanics, and to differential geometry.

Path integration is one of the most powerful methods in modern theoretical physics. A standard technique for investigating a new model in quantum field or string theories is to sum over everything with what one hopes are the correct weights. Indeed some would regard the generic expression $\int Dx \exp(-iS)$ as the fundamental route to quantisation. Certainly formal manipulation of path integrals gives extremely valuable insight into a theory, but ultimately the true meaning of the path integrals used should be established. This is particularly true in the case of path integrals for fermions, where the integrals are not even the limits of sums, and standard measure theoretical results do not apply. For instance, it is usually assumed that the sum over paths can be replaced by a sum over Fourier modes, which is far from obvious when the word sum is being used in a formal sense. (This procedure is justified in these lectures, and the correct normalisation derived.) In the simplest situation, bosonic quantum mechanics in imaginary time, there is a rigorous theory, based on ideas of Wiener, Feynman and Kac, and described fully in Simon's excellent book "Functional integration and quantum physics" [1]. These lectures aim to describe how a similar theory may be developed for purely fermionic systems and for mixed fermi-bose systems in quantum mechanics.

The first lecture considers purely fermionic systems. First an extension of the Berezin method of integration to the very infinite dimensional space of paths in anticommuting space is developed; a fermionic analogue of Wiener measure is then defined, and used to derive a Feynman-Kac formula for the kernel of the evolution operator. In the second lecture these fermionic path integrals are combined with conventional path integrals; also a supersymmetric square root of Wiener measure is described, and used to give a simple Feynman-Kac formula for supersymmetric systems. In the third and final lecture these methods are extended to curved space, and applied to give a rigorous version of the supersymmetric proof of the Gauss-Bonnet theorem.

1. FERMIONIC PATH INTEGRATION

In this lecture an analytic theory of fermionic path integration is developed, in a manner as closely analoguous as possible to the bosonic method. An important step is the construction of an integration theory for spaces of anticommuting variables of uncountably infinite dimension which reduces to Berezin integration in the finite dimensional case, in a manner parallel to Bochner's treatment of conventional stochastic processes [2], where the primary objects are the finite dimensional marginal distributions. Generally, when anticommuting variables are used in fermionic quantum mechanics, only finite dimensional linear algebra is involved. In this case, because of the presence of the continuous time parameter, some genuine analysis is required.

In section 1.1 bosonic path integration is reviewed, with emphasis on those features which will be generalised in the fermionic approach. Section 1.2 contains a description of the "Schrödinger" representaion for fermionic quantum mechanics on spaces of functions of anticommuting variables, while further details of such functions are described in 1.3. In section 1.4 Berezin integration is extended to infinite dimensional anticommuting space, and Grassmann "Wiener" measure and Brownian motion are introduced. Section 1.5 contains a proof of the Feynman-Kac formula for fermions, that is the expression of the kernel of the evoltion operator exp(-Ht) in terms of a path integral using the Wiener measure developed in the previous section. In section 1.6 Grassmann Brownian motion is expressed as a Fourier sum of independent (Grassmann) Gaussians, thus making rigorous the heuristic physicist's replacement of the sum over paths by the sum over Fourier modes with anticommuting boundary conditions [3].

There is a considerable literature on various non-commutative versions of probability theory [4,5]; however these theories do not seem to be directly applicable to the type of path integrals considered in this paper. The closest is the work of Haba [5] but the anticommuting Brownian motion constructed there is in position space, not phase space, and would not lead to the Feynman-Kac formula presented here.

1.1 A brief description of bosonic path integration

In the canonical quantisation of n-dimensional particle mechanics, the classical observables p^i (momentum) and x^i (position) $(i = 1,\dots,n)$ are replaced by the quantum operators $\hat{p}^i$ and $\hat{x}^i$; also the classical Poisson brackets

$$\{x^i, p^j\} = \delta^{ij} \tag{1.1.1}$$

are replaced by the quantum commutation relations

$$[\hat{x}^i, \hat{p}^j] = i\hbar\delta^{ij}. \tag{1.1.2}$$

The standard representation is the Schrödinger representation, where $\hat{p}^i$ and $\hat{x}^i$ are represented on the space $L^2(R^n)$ of square integrable complex valued functions of R^n by

$$\hat{p}^i f(x) = -i(\partial/\partial x^i) f(x) \tag{1.1.3}$$

$$\hat{x}^i f(x) = x^i f(x). \tag{1.1.4}$$

(As usual, units are used in which Plank's constant $\hbar$ is set to one.)

The evolution of the system is determined by the Schrödinger equation

$$\partial f/\partial t = -Hf, \tag{1.1.5}$$

where H is the Hamiltonian of the system; the system is solved if one knows the evolution operator $\exp(-Ht)$; path integration provides a method for calculating the kernel of this operator.

In physicist's notation, if $H = \hat{p}^2 + V(\hat{x})$ is the Hamiltonian, the key formula is

$$\langle q|\exp(-Ht)|q'\rangle = \int Dx \exp(-\int_0^t \{\dot{x}^2 + V(x)\}dt), \tag{1.1.6}$$

where the integration is over all paths beginning at q and ending at q'.The corresponding analytic formula is [1]

$$\exp(-Ht)(q,q') = \int d\mu(q,q',t) \exp(-\int_0^t V(x)\, dt), \tag{1.1.7}$$

where μ is conditional Wiener measure on the space of paths in R. By including the kinetic term $\dot{x}^2$ in the measure rather than in the integrand a true measure, free from infinite dimensional normalisation factors and so on, is obtained.

There are two ways of defining this measure. One approach is to define the measure of certain measurable sets, and to use a limiting process to obtain the measure of more general sets. An alternative approach, which will be described in more detail because it seems to be the appropriate one to generalise to the anticommuting case, uses Kolmogorov's extension theorem. The rough idea of this is as follows. Suppose that $\{\Omega,F,\mu\}$ is a probability measure space, and $\{f(t)|t\in I\}$ are random variables (I being an arbitary set). Also, suppose that for each finite $J = \{i_1,\ldots,i_N\}\subset I$, μ_J denotes the joint distribution of $f=(f(t_1),\ldots,f(t_N))$, that is, μ_J is the probability measure on R^N defined by

$$\mu_J(A) = \mu(f^{-1}(A)), \tag{1.1.8}$$

A being a measurable subset of R^N. Then, if $J'=\{t_1,\ldots,t_M\}$, with M>N,

$$\mu_{J'}(A \times R^{M-N}) = \mu_J(A). \tag{1.1.9}$$

Conversely, the Kolmogorov extension theorem asserts that if one strarts with a set I and, corresponding to each finite subset $J=\{t_1,\ldots,t_N\}$ of I, a probability measure μ_J on R^N, such that whenever $J\subset J'\subset I$, with $J=\{t_1,\ldots,t_N\}$ and $J'=\{t_1,\ldots,t_M\}$, one has

$$\mu_{J'}(A \times R^{M-N}) = \mu_J(A), \tag{1.1.10}$$

then there exists a probability space $\{\Omega,F,\mu\}$ and random variables $\{f(t)\,|\,t\in I\}$ such that for each finite subset $J=\{t_1,\ldots,t_N\}$ of I the μ_J are the joint distributions of $f = (f(t_1),\ldots,f(t_N))$. The usual model has $\Omega = R^I$, with f(t) denoted x(t).

Conditional Wiener measure (for paths beginning at a and ending at b) may be defined in this way by letting I=(0,t) and, given $J=\{t_1,\ldots,t_N\}$ with $0<t_1<\ldots<t_N<t$, setting

$$\mu_J = P(t_1,a,x_1)P(t_2-t_1,x_1,x_2)\ldots P(t-t_N,x_N,b)\; dx^1\ldots dx^N \tag{1.1.11}$$

where $$P(t,x,y) = (2\pi t)^{-\frac{1}{2}} \exp(-(x-y)^2/2t). \tag{1.1.12}$$

(This corresponds to the kinetic term contribution in the physicist's more heuristic approach. P(t,x,y) is simply the kernel of the operator $\exp(-t\hat{p}^2)$, $\hat{p}^2$ being the free Hamiltonian.) The finite distributions μ_J defined in this way satisfy the Kolmogorov consistency condition (1.1.10) because of the semigroup property of $\exp(-t\hat{p}^2)$. The Feynman-Kac formula (1.1.7) can then be proved by a limiting process, as described in [1].

1.2 Fermionic quantum mechanics

In order to develop an analogous treatment of fermionic path integration, a "Schrödinger" representation for fermionic quantum mechanics is required. Suppose that one has n fermi operators Ψ^i obeying canonical anticommutation relations

$$\{\Psi^i,\Psi^j\} = \delta^{ij}. \tag{1.2.1}$$

These can be represented by the operators

$$\Psi^i = \theta^i + \partial/\partial\theta^i \tag{1.2.2}$$

acting on the space $H^\infty(B_L^{0,n},C)$ of polynomial functions of n anticommuting variables. (Further details of these functions will be described in the next sub-section.) Other representations of (1.2.1) using anticommuting variables are of course possible, and would lead to similar theories of path integration to that described here; the representaion used here is appropriate for the proof of the Gauss-Bonnet theorem discussed in section 3. The use of anticommuting variables to represent anticommutation relations, and to develop a fermionic version of the Feynman principle, was first demonstrated by Martin [6].

The dynamics of such a fermionic quantum mechanical system are determined by a Hamiltonian $H = V(\Psi)$, with corresponding Schrödinger equation

$$\partial f/\partial t = -Hf \tag{1.2.3}$$

where the wave function f is a function of time t and n anticommuting variables. The aim of fermionic path integration is to construct a "measure" $d\mu$ on the space of paths in anticommuting space so that the kernel of the evolution operator exp(-Ht) may be expressed in terms of a Feynman-Kac formula. Before looking at a method for constructing such a measure, an aside on Grassmann variables, and analysis of functions of Grassmann variables, will be given.

Various formulations of fermionic quantum mechanics in terms of Grassmann variables exist in the literature (in addition to the work by Martin cited above) [7].

1.3 Analysis of functions of Grassmann variables

For each positive integer L, B_L will denote $\Lambda(R^L)$, the real

Grassmann algebra with L anticommuting generators, while C_L will denote $\Lambda(C^L)$. The actual value of L will not be important, except that it will always be assumed to be at least twice the number of anticommuting variables in the space under consideration. (It can be shown that the natural injection $B_L \to B_{L+1}$ can always extend to superspaces, even curved, constructed from B_L, and a direct limit considered so that one is effectively using an infite number of Grassmann generators [8].) $B_{L,0}$ and $B_{L,1}$ will denote respectively the even and odd parts of B_L, while $B_L^{m,n}$ will denote the cartesian product of R^m with n copies of $B_{L,1}$. (In the purely fermionic case, m will always be zero.) If I is an arbitrary set $B_L^{(m,n)I}$ will denote the cartesian product $\prod_{t\in I} B_L^{m,n}(t)$ where each Grassmann algebra $B_L(t)$ is distinct. (The purpose of using this product rather than a straightforward product of copies of $B_L^{m,n}$ is to avoid unwanted cancellations caused by repeated Grassmann generators.)

One of the function spaces used is $H^\infty(B_L^{0,n},C)$, the set of functions $f:B_L^{0,n} \to B_L$ such that

$$f(\theta^1,\dots,\theta^n) = \sum_{\mu\in M_n} f_\mu\, \theta^\mu \tag{1.3.1}$$

where $\mu = \mu_1\dots\mu_k$ is a multiindex with $1\leqq\mu_1<\dots<\mu_k\leqq n$, and M_n is the set of all such multiindices, including the empty index denoted $\emptyset$. Also, the coefficients f_μ are complex numbers, and θ^μ denotes the product $\theta^{\mu_1}\dots\theta^{\mu_k}$. (All roman indices will be single indices and all greek indices will be multiindices; a summation convention will be used where repeated roman indices will be summed over their range, and repeated greek indices summed over M_n for the appropriate value of n.)

If K is an operator on $H^{\infty}(B_L{}^{o,n},C)$, then

$[K]:B_L{}^{o,2n} \to B_L$ is said to be its kernel if, given f in $H^{\infty}(B_L{}^{o,n},C)$,

$$Kf(\theta^1,\ldots,\theta^n) = \int d^n\phi\, [K](\theta^1,\ldots\theta^n,\phi^1,\ldots\phi^n)\, f(\phi^1,\ldots,\phi^n), \tag{1.3.2}$$

the integral with respect to the Grassmann variable ϕ being the usual Berezin integral [8]. If I is the identity operator on $H^{\infty}(B_L{}^{o,n},C)$, then

$$[I](\theta,\phi) = \delta(\theta,\phi) = (\theta^1-\phi^1)\ldots(\theta^n-\phi^n)$$
$$= \int d^n\kappa\, \exp(-i\kappa_i(\theta^i-\phi^i)). \tag{1.3.4}$$

Also, if $V(\Psi) = V_\mu\, \Psi^\mu$ (with Ψ the operators defined in (1.2.2)above), then

$$V(\Psi)\ (\theta,\phi) = \int d^n\kappa\, V(\theta+i\kappa)\, \exp(-i\kappa_i(\theta^i-\phi^i)). \tag{1.3.5}$$

Now $H^{\infty}(B_L{}^{o,n},C)$ has the structure of a 2^n-dimensional complex vector space, and operators which have kernels are precisely linear operators on this space. Moreover it can be shown that, if K is a linear operator on $H^{\infty}(B_L{}^{o,n},C)$, then

$$\text{trace } K = \int d^n\theta\, [K](\theta,-\theta). \tag{1.3.6}$$

In the second and third lectures further classes of functions of $B_L{}^{r,s}$ will be required. A function $f:B_L{}^{r,s} \to B_L$ will be said to be H^{∞} if, for each μ in M_s, there exists a C^{∞} function $f_\mu:R^r \to R$ such that

$$f(x,\theta) = f_\mu(x)\theta^\mu. \tag{1.3.7}$$

The space of all such H^{∞} functions will be denoted $H^{\infty}(B_L{}^{r,s},R)$. In a similar way $H^{\infty}(B_L{}^{r,s},C)$ is defined to be the space of all functions of $B_L^{r,s}$ into C_L expressible in the form (1.3.7) with the coefficient functions f_μ required to be C^∞ functions of R^r into C. More generally, if A is a finite-dimensional algebra over the reals, $H^\infty(B_L^{r,s},A)$ denotes the space of functions f of $B_L^{r,s}$ into A x B_L satisfying

$$f(x,\theta) = f_\mu \times \theta^\mu \tag{1.3.8}$$

where the coefficient functions f_μ are C^∞ functions of R^r into A.

Even derivatives ∂_i and odd derivatives δ_i of a H^∞ function are defined by

$$\partial_i f(x^1,\ldots,x^r;\theta^1,\ldots,\theta^s) = \partial_i f_\mu(x^1,\ldots,x^r)\theta^\mu$$

$$\text{for } i = 1,\ldots,r$$

and $\quad \delta_j f_\mu(x^1,\ldots,x^r)\theta^\mu \quad$ (no summation)

$$= -(-1)^p f_\mu(x^1,\ldots,x^r)\theta^{\mu_1}\ldots\theta^{\mu_{p-1}}\theta^{\mu_{p+1}}\ldots\theta^{\mu} \quad \text{if } j = \mu_p$$

$$= 0 \text{ otherwise} \quad \text{for } j = 1,\ldots,s. \tag{1.3.9}$$

Other spaces of functions (such as $L^{2,}(B_L^{r,s},C)$ and $C_0^{\infty,}(B_L^{r,s},R)$ will be defined in a similar way by imposing the appropriate restrictions on the coefficient functions f_μ. A norm $\|\cdot\|$ will be defined on $L^{2,}(B_L^{r,s},C)$ by setting

$$\| f \| = \sum_{\mu \in M_s} \| f_\mu \| . \tag{1.3.10}$$

With this norm $L^{2\prime}(B_L{}^{r,s},C)$ becomes aHilbertspace.

Also $L^{2\prime}(B_L{}^{r,s},C)$ is the closure of $H^{\infty}(B_L{}^{r,s},C)$ with $L^{2\prime}$ norm.

1.4 Extending the Berezin integral to infinite-dimensional spaces.

The conventional theory of integration is based on the concept of measurable sets, and as such is quite inapproprate for generalisation to the Berezin case, where limits of integration are not used. The approach taken here is to invert the Kolmogorov consistency conditions described in 1.1, and to use them as a definition of measure on $B_L{}^{(0,n)I}$, along the lines of Bochner's approach to conventional integration[2].

Definition 1.4.1

An n-Grassmann probability space of weight $w \in C_L$ consists of

(a) a set I;

(b) for each finite subset J of I a function $f_J \in H^{\infty}(B_L{}^{(0,n)J},B_L)$

such that

(i) $\int d^{n|J|}\theta\, f_J(\theta) = w.$ (1.4.1)

(ii) if $J = \{b_1,\ldots,b_s\}$ and $J' = \{b_1,\ldots,b_{s-1}\}$ are subsets of I, then

$$f_{J'}(\theta^{i1},\ldots,\theta^{is-1}) = \int d^n\theta^s\, f_J(\theta^{i1},\ldots,\theta^{is}). \tag{1.4.2}$$

Such a space will be denoted $(B_L{}^{(0,n)A},\{f_B\},d\mu)$.

An example of a 2n-Grassmann probability space is given by what will be called Grassmann Wiener measure with parameters t,α,β,κ (where t is a positive real number and α,β and κ are in $B_L{}^{0,n}$). In this case $I=(0,t)$ and, given $J=\{t_1,\ldots,t_N\}$ with $0<t_1<\ldots<t_N<t$,

$f_J(\theta^1,\kappa_1,\ldots,\theta^N,\kappa_N)$

$$= \exp i(-\kappa_i(\theta^{i1}-\alpha^i)-\rho_{i1}(\theta^{i2}-\theta^{i1})\ldots-\rho_{iN}(\beta^i-\mu^{iN})).$$

(1.4.3)

Grassmann Wiener measure with parameters t,α,β and κ will be denoted $d\mu[t,\alpha,\beta,\kappa]$. As with bosonic Wiener measure, the kernel of the free Hamiltonian (in this case the delta function) has been used to define the measure. The phase-space version is used because it leads to a simple Feynman-Kac formula. It is straightforward to check that the functions f_J of equation (1.4.3) do satisfy the consistency conditions (1.4.2). The weight of the measure is $\exp(-i\kappa_i(\alpha^i-\beta^i))$. The extra parameter κ (in addition to the parameters α and β which one might expect by analogy with conventional conditional Wiener measure) arises because one is working in phase space.

Having defined measure on $B_L^{(0,n)I}$, the next step is to consider what can be integrated with respect to this measure, that is, to define a notion of random variable, and specify how the integral of a random variable may be calculated.

Definition 1.4.2

A Grassmann random variable g on an n-Grassmann probability space $(B_L^{(0,n)I},\{f_J\},d\mu)$ consists of

(i) asequence$J_1,J_2,\ldots$ of finite subsets ofI.

(ii) a sequence of functions $g_1,g_2,\ldots$ with $g_r \in H^\infty(B_L^{(0,n)J},B_L)$ for each $r = 1,2,\ldots$ such that $\int d^{n|J|}\theta\, f_J(\theta)\, g_r(\theta)$ tends to a limit as r tends to infinity

This limit will be denoted $\int d\mu\, g$ and called the expectation of g.

The idea of this definition is that, as well as allowing one to integrate simple functions such as $\theta(t_k)$ on Grassmann Wiener space, it is also possible to integrate some functions which depend (loosely speaking) on an infinite number of t's. An example is

$$\exp(-\int_0^t V(\theta(s)+i\rho(s))ds),$$

the integrand in the Feynman-Kac formula to be proved in the next sub-section. This random variable will now be properly defined, and shown actually to be a random variable.

Proposition 1.4.3

Suppose that $V \in H^{\infty}(B_L{}^{0,n}, C)$.

For r= 1,2... let J_r denote the subset $\{jt/2^r \mid j=1,\ldots,2^r-1\}$,

and let $g_r \in H^{\infty}(B_L{}^{(0,n)J}, B_L)$ be defined by

$$g_r(\theta,\rho) = \exp - \sum_{j=1}^{2^r-1} (t/2^r)\, V(\theta^j + i\rho_j). \tag{1.4.4}$$

Then the sequences J_r, g_r define a random variable on Grassmann Wiener space with parameters t, α, β, κ.

Outline of proof: (a different proof may be found in [10])

Let $S_r = \int d\theta\, d\rho\, f_J(\theta,\rho) g_r(\theta,\rho)$.

Then $S_r - S_{r-1}$

$$= \left(\exp - \sum_{j=1}^{2^r-1} (t/2^{r+1}) V(\theta^{2j} + i\rho^{2j})\right)$$

$$\times \left\{\exp - \sum_{j=1}^{2^r-1} (t/2^{r+1}) V(\theta^{2j+1} + i\rho^{2j+1}) \right.$$

$$\left. - \exp - \sum_{j=1}^{2^r-1} (t/2^r) V(\theta^{2j} + i\rho^{2j}), \right. \tag{1.4.5}$$

and hence it can be shown that $|S_r - S_{r-1}| < K(t/2^r)^2 \times 2^r$, for some positive constant K. (This estimate is a result of the particular measure used; essentially it follows because the measure involves differences $\theta^i - \theta^{i-1}$ and $\rho_i - \rho_{i-1}$.) Thus the sequence S_r tends to a limit as r tends to infinity, as required.

The random variable defined in proposition 1.4.3 will be denoted

1.5 The fermionic Feynman-Kac formula

Having defined Grassmann Wiener measure, it is now possible to prove the Feynman-Kac formula for the kernel of the evolution operator of a fermionic quantum mechanical system.

Theorem 1.5.1

Suppose that V is a function in $H^{\infty}(B_L^{0,n},C)$, that is,

$$V(\theta) = V_{\mu}\, \theta^{\mu}, \tag{1.5.1}$$

and that H is the operator $V(\Psi)$ on $H^{\infty}(B_L^{0,n},C)$. Then

$$\exp(-Ht)(\alpha,\beta) = \int d^n\kappa\, d\mu[t,\alpha,\beta,\kappa]\; \exp\left(-\int_0^t V(\theta(s)+i\rho(s))\, ds\right). \tag{1.5.2}$$

Proof. First suppose that, for r=1,2,...,N, f_r is a function in $H^{\infty}(B_L^{0,n},C)$. Then it can be shown by repeated use of (1.3.5) that

$$f_1(\Psi)\ldots f_N(\Phi)\, (\alpha,\beta)$$
$$= \int d^n\kappa\, d\mu\, f_1(\theta(t_1)+i\rho(t_1))\ldots f_N(\theta(t_N)+i\rho(t_N)), \tag{1.5.3}$$

where $0<t_1<\ldots<t_N<t$. Secondly,

$$[1-(t/2^r)V(\Psi)](\alpha,\beta)$$
$$= \int d^n\rho\, \exp(-i\rho_i(\alpha^i-\beta^i)-(t/2^r)V(\alpha+i\rho)) \;+ o(t/2^r), \tag{1.5.4}$$

and hence, using the compound interest formula

$$(1-k^2/N + o(1/N))^N = \exp(-k^2) + o[1], \tag{1.5.5}$$

$$[\exp(-tH)](\alpha,\beta)$$

$$= \lim_{N\to\infty} \int d^n\kappa \, d^{2n}\theta \, d^{2n}\rho$$

$$\exp-\{ \sum_{s=1}^{2^N-1} (-(t/2^N)V(\theta^s+i\rho_s))-i\kappa_1(\theta^{1i}-\alpha)$$

$$-i\rho_{1i}(\theta^{2i}-\theta^{1i})\ldots-i\rho_{Ni}(\beta^i-\theta^{Ni})\}$$

$$= \int d^n\kappa \, d\mu[t,\alpha,\beta,\kappa] \exp -\int_0^t V(\theta(s)+i\rho(s))ds.$$

(1.5.6)

1.6 The fourier mode expansion of the fermionic Brownian bridge.

By analogy with terminology for conventional Brownian motion, the process $\Psi(s)=(\theta(s)+i\rho(s))$ corresponding to Grassmann Wiener measure with parameters 1,0,0,0 will be called the fermionic Brownian bridge. Also a Grassmann process will be called Gaussian if its finite distributions, (that is, the f_J of section 1.4) are all exponentials of quadratic functions. As with conventional Gaussian processes, it can be shown that Grassmann gaussian processes are entirely characterised by their covariances [10].

Now it can be shown by direct calculation that the fermionic Brownian bridge is Gaussian, with (given $0<s,t<1$)

$$\mathrm{cov}(\Psi^i(s),\Psi^j(t)) = \int d\mu \, \Psi^i(s) \, \Psi^j(t)$$

$$\begin{cases} = -i\,\delta^{ij} & \text{if } s<t \\ = 0 & \text{if } s=t \\ = +i\,\delta^{ij} & \text{if } s>t. \end{cases}$$

(1.6.1)

If, for $i = 1,\ldots,n$ and $r = 1,2,\ldots,N$. X^{ir}, Y_{ir} are Gaussian Grassmann random variables with joint distributions

$$\prod_{r=1}^{N} d^n\theta^r \, d^n\rho^r \exp(-i\rho_{ir}\theta^{ir})$$

then, in the limit as N tends to infinity, the n-dimensional Grassmann Gaussian process

$$w^i(t) = \sum_{r=1}^{N} (2/\sqrt{(2r+1)\pi})\{\sin(2r+1)\pi t\, X^{ir} + i\cos(2r+1)\pi t\, Y_{ir}\} \tag{1.6.2}$$

has the same covariances as $\Psi^i(t)$, and hence is a version of $\Psi^i(t)$. (The concept of version is a technical one, borrowed from conventional statistics, and defined in [10]; the idea is that to all intents and purposes the two processes $\Psi(t)$ and $w(t)$ are interchangeable.) For some applications it is easier to use the version 1.6.2 of fermionic Brownian motion; an example is the calculation of the heat kernel of the two-dimensional fermionic oscillator, defined by the Hamiltonian

$$H = ik\Psi^1\Psi^2, \tag{1.6.3}$$

using the Feynman-Kac formula (1.5.2), as described in [10].

2. SUPERSYMMETRIC PATH INTEGRATION

An alternative title for this lecture could be "the square root of everything". In the first section, on supersymmetric quantum mechanics, the square root of the Schrödinger equation is described. Next a "super" stochastic process (partly bosonic and partly fermionic) which might be described as the square root of the Wiener process is considered; the third section contains a Feynman-Kac formula which uses the super process just constructed and the square root of the Hamiltonian. In the course of proving this formula, a square root of the fundamental theorem of calculus is used.

In supersymmetric quantum mechanics with a single Hermitian super charge Q, the Hamiltonian H is the square of Q. The aim of this lecture is to show that , if Q has the form

$$Q = (\hat{\Psi}.\hat{p})/\sqrt{2} + W(\hat{x},\hat{\Psi}) \tag{2.1}$$

($\hat{\Psi}$ being the fermi operator of (1.2.2)), then the kernel of the super evolution operator exp(-Ht-Qτ) [11], where τ is an anticommuting "time", can be evaluated by integration in the space of superpaths

$\{x(t,\tau),\Psi(t,t)\}$ in superspace according to the formula

$$\exp(-Ht-Q\tau)(a,\alpha,b,\beta)$$

$$= \int Dx\, D\Psi \exp -\int d\sigma \int_0^{t+\sigma\tau} ds\ \partial x(s,\sigma)/\partial s \cdot \Psi(s,\sigma) + W(x(s,\sigma),\Psi(s,\sigma)) \tag{2.2}$$

(Here α and β are anticommuting parameters, and the integration variables s and σ are respectively commuting and anticommuting.) Of course (2.2) is a heuristic version of the result (analogous to (1.1.6)), and a supersymmetric generalisation of Wiener measure will be required to give a rigorous version of the result.

This result is useful for computation because the superpotential W is generally much simpler than the potential V. (An example is given at the end of the paper.) The result (2.2) is also interesting because it promotes Q to a fundamental role in the path integration of supersymmetric theories, which seems appropriate because in such theories Q is more fundamental than H. In other situations, such as supersymmetric field theory or string theory, the analogue of the quantity $x \cdot \Psi + W(x,\Psi)$, which occurs in the exponent in the integrand of (2.2), may prove significant. This quantity is not the square root of the Lagrangian, but relates to the supercharge Q rather as the Lagrangian relates to the Hamiltonian H.

It is also shown in this lecture that $x(t,\tau)$ and $\Psi(t,\tau)$ may be expressed in terms of the superpath $X(t,\tau) = x(t)+i\tau\Psi(t)/\surd 2$, where $x(t)$ is an ordinary bosonic path or Brownian motion and $\Psi(t)$ is the fermionic equivalent described in lecture 1.

The level of mathematical rigour in this lecture is somewhat lower than that of lecture 1; many of the ideas have still to be developed properly, but the shape of them seems sufficiently exciting to warrant inclusion at this stage.

2.1 Supersymmetric quantum mechanics

Consider a system with 2n bose operators $\hat{x}^i, \hat{p}^i$ (i=1,...,n) and n fermi operators $\hat{\Psi}^i$, and Hamiltonian

$$H = \hat{p}^2/2 + V(\hat{x},\hat{\Psi}) \tag{2.1.1}$$

where

$$H = Q^2 \qquad (2.1.2)$$

with Q a Hermitian operator of the form

$$Q = (\hat{\Psi}^i \hat{p}_i)/\sqrt{2} + W(\hat{x},\hat{\Psi}). \qquad (2.1.3)$$

Systems of this form, with H the square of an operator Q, are called supersymmetric, and the operator Q is called the supercharge. H_0 and Q_0 will denote the free Hamiltonian and supercharge (obtained by setting $W(\hat{x},\hat{\Psi})$ and hence $V(\hat{x},\hat{\Psi})$ to zero).

As usual, the evolution of the system is determined by the Schrödinger equation

$$df/dt = -Hf. \qquad (2.1.4)$$

In supersymmetric theories, by introducing an anticommuting time τ in addition to ordinary (imaginary) time t, one may construct a square root of the Schrödinger equation [11]. This is the equation

$$D_{t,\tau}\, f = Qf \qquad (2.1.5)$$

where

$$D_{t,\tau} =_{def} \partial/\partial\tau + \tau\partial/\partial t \qquad (2.1.6)$$

and f is a function of x^i,θ^i $(i=1,\ldots,n)$, t and τ. The solution to this equation with initial condition

$$f(x,\theta,t,\tau) = g(x,\theta) \text{ when } t=0,\tau=0 \qquad (2.1.7)$$

is given by

$$f = \exp(-Ht-Q\tau)g. \qquad (2.1.8)$$

In the rest of this lecture a measure in the space of superpaths in superspace will be developed, so that the kernel of the super evolution operator $\exp(-Ht-Q\tau)$ can be expressed as an integral with respect to this measure, corresponding to the heuristic formula (2.2).

2.2 The "square root" of the Wiener process

Just as the purely Bosonic path integral is based on the kernel of $\exp(-H_0 t)$ the supersymmetric path integral will be based on the kernel of $\exp(-H_0 t - Q_0\ \tau)$. This operator has kernel

$$[\exp(-H_0 t - Q_0 \tau)](a,b,\alpha,\theta) = \int d^n\kappa\, P(a,b,\alpha,\beta,\kappa,t,\tau) \tag{2.2.1}$$

where the variables $\kappa^1,\dots,\kappa^n$ are anticommuting and

$$P(a,b,\alpha,\beta,\kappa,t,\tau)$$

$$=_{\text{def}} (2\pi t)^{-n/2}\exp\{-(b-a-(i/\sqrt{2})\tau(\beta+i\kappa))^2/2t - i\kappa_i(\alpha^i-\beta^i)\}. \tag{2.2.2}$$

(Here it should be explained that the concept of kernel for operators on superspace function spaces is a combination of the conventional idea of kernel for functions of real variables and the kernel introduced in lecture 1 for functions of anticommuting variables. In more detail, suppose that $K:L^{2'}(B_L^{r,s},C) \to L^{2'}(B_L^{r,s},C)$. Then, if there exists a function

$$[K] : B_L^{r,s} \times B_L^{r,s} \to C_L$$

polynomial in anticommuting arguments such that, for all f in $L^{2'}(B_L^{r,s})$ and all (x,θ) in $B_L^{r,s}$, the integral $\int d^n y d^n\phi\, [K](x,y,\theta,\phi)\, f(y,\phi)$ exists and satisfies

$$K(f)(x,\theta) = \int d^n y\ d^n\phi\, [K](x,y,\theta,\phi)\, f(y,\phi), \tag{2.2.3}$$

the function $[K]$ is said to be the integral kernel of the operator K. (The integration over the anticommuting variables ϕ_j in (2.2.3) is carried out according to the usual Berezin prescription [9].) Kernels of operators on other spaces of functions of $B_L^{r,s}$ may be defined similarly when appropriate.

Equipped with the kernel (2.2.1) of the free super evolution operator, it seems natural to define a "measure"

$d\mu[a,b,\alpha,\theta,\mu,t,\tau]$ on the space $B_L^{(n,n)I'}$ (where I' $=I \times B_L^{(0,1)}$) of superpaths $(x(t,\tau),\Psi(t,\tau))$ in superspace by finite distributions

$$\begin{aligned} f_J &= \{P(a,x_1,\alpha,\theta_1,\kappa,t_1,\tau\,) \\ &\times P(x_1,x_2,\theta_1,\theta_2,\rho_1,t_2-t_1+\tau_2\tau_1,\tau_2-\tau_1) \times \ldots \\ &\times P(x_{N-1},x_N,\theta_{N-1},\theta_N,\rho_{N-1},t_N-t_{N-1}+\tau_N t_{N-1},\tau_N-t_{N-1}) \\ &\times P(x_N,b,\theta_N,\beta,\rho_N,t-t_N+\tau\tau_N,\tau-\tau_N) \end{aligned} \tag{2.2.4}$$

where $J = (t_1,\tau\,,\ldots,t_N,\tau\,)$ and $0<t_1<\ldots<t_N<t$.

To make sense of these finite distributions, a proper definition of measure and random variable on infinite dimensional superspace is required

The approach taken to measures on $B_L^{(r,s)A}$ will be a combination of conventional measures on $(R^r)^A$ (in the Bochner approach, where finite dimensional marginals are the primary objects[2]) with the Berezin type measures on $B_L^{(0,s)A}$ described in lecture 1, leading to following definition of super probability measure on $B_L^{(r,s)A}$.

Definition 2.2.1 An (r,s)-super probability space of weight $w \in C_L$ consists of

(a) a set A

(b) for each finite subset $J=\{t_1,\ldots,t_k\}$ of A a function

f_J $L^{2,}(B_L^{(r,s)J},C_L)$ such that

(i) $\int d^{rk}x\, d^{sk}\theta\, f_J(x,\theta) = w$

(ii) if $J' = \{t_1,\ldots,t_{k-1}\}$ then

$$f_{J'}(x^1,\theta^1,\ldots x^{k-1},\theta^{k-1}) = \int d^r x^k\, d^s\theta^k\, f_J(x^1,\theta^1,\ldots,x^k,\theta^k)\,. \tag{2.2.5}$$

Such a super probability space will be denoted $\{B_L^{(r,s)A},\{f_J\},d\mu\}$. The f_J will be referred to as the finite distributions of the super measure $d\mu$.

The aim of this approach is to build Kolmogorov-type consistency conditions into the definition and thus be able to integrate certain objects, not quite functions of $B_L^{(r,s)A}$, which will be called super random variables.

Definition 2.2.2 A super random variable g on an (r,s)-super probability space $\{B_L^{(r,s)A},\{f_J\},d\mu\}$ consists of

(i) a sequence $J[1],J[2],\ldots$ of finite subsets of A

(ii) for each positive integer N afunction g_N in $L^{2\prime}(B_L^{(r,s)J(N)},C_L)$ such that (if $|J(N)|$ denotesthe number of elements in the set J(N)) the sequence $I_1,I_2,\ldots$ with

$$I_N =_{def} \int d^{r|J(N)|}x\, d^{s|J(N)|}\theta\, f_{J(N)}(x,\theta)g_N(x,\theta) \tag{2.2.6}$$

tends to a limit as N tends to infinity. This limit will be denoted $\int d\mu\, g$.

The distributions defined by equation (2.2.4) do satisfy the necessary consistency conditions, since

$$\exp(-Hs-Q\sigma)\exp(-Ht-Q\tau) = \exp(-H(s+t+\sigma t) - Q(\sigma+\tau)). \tag{2.2.7}$$

2.3 The supersymmetric Feynman-Kac formula

In terms of the measure on superpaths which has just been defined, the key result (2.2) may be expressed as

$$\begin{aligned}&[\exp(-Ht-Q\tau)](a,b,\alpha,\beta)\\ &= \int d^n\kappa\, d\mu[a,b,\alpha,\beta,\kappa,t,\tau]\exp\{-\int d\sigma \int_0^{t+\sigma\tau} ds\, W(x(s,\sigma),\Psi(s,\sigma))\}.\end{aligned} \tag{2.3.1}$$

This result is not easily proved by direct assault, but may be proved by an extension of a method developed by Guerra, Rosen and Simon for proving the bosonic Feynman-Kac formula [12]. Three formulae from elementary (1,1) dimensional supercalculus are required: first, the fundamental theorem of supercalculus,

$$\int d\sigma \int_0^{t+\sigma\tau} ds \; D_{s,\sigma} f(s,\sigma) = f(t,\tau) - f(0,0),$$

(2.3.2)

is required. Secondly, there is the formula for differentiation with respect to the limits of integration,

$$D_{t,\tau} \left(\int d\sigma \int_0^{t+\sigma\tau} ds \; f(s,\sigma) \right) = f(t,\tau),$$

(2.3.3)

and thirdly there is the formula for change of variable

$$\int d\sigma \int_{u+\sigma\eta}^{t+\sigma\tau} ds \; f(s,\sigma) = \int d\sigma \int_0^{t-u+\sigma(\tau-\eta)+\eta\tau} ds \; f(s+u+\eta\sigma, \sigma+\eta).$$

Combining (2.3.2) and (2.3.3) one may deduce that

$$\exp\{-\int d\sigma \int_0^{t+\sigma\tau} ds \; g(s,\sigma)\}$$

$$= 1 - \int d\eta \int_0^{t+\eta\tau} du \; \{(g(u,\eta) \exp(-\int d\sigma \int_{u+\sigma\eta}^{t+\sigma\tau} ds \; g(s,\sigma))\}.$$

(2.3.5)

Now let $B_{t,\tau}$ be the operator with kernel

$$[B_{t,\tau}](a,b,\alpha,\beta)$$

$$= \int d^n\kappa \int d\mu[a,b,\alpha,\beta,\kappa,t,\tau] \exp - \int d\sigma \int_0^{t+\sigma\tau} ds \; W(x(s,\sigma),\Psi(s,\sigma)).$$

(2.3.6)

Taking (2.3.5) with $g(s,\sigma) = W(x(s,\sigma),\Psi(s,\sigma))$, changing variable according to (2.3.4), and integrating with measure $d^n\kappa d\mu[a,b,\alpha,\beta,\kappa,t,\tau]$ gives

$$[B_{t,\tau}](a,b,\alpha,\beta) = [\exp(-H_0t-Q_0\tau)](a,b,\alpha,\beta)$$

$$- \int d\eta \int_0^{t+\eta\tau} du\, [\exp-(H_0u+Q_0\eta)\ W\ B_{t-u+\tau\eta,\tau-\eta}](a,b,\alpha,\beta). \qquad (2.3.7)$$

A further change of variables gives

$$B_{t,\tau}(a,b,\alpha,\beta) = [\exp(-H_0t-Q_0\tau)](a,b,\alpha,\beta)$$

$$- \int d\sigma \int_{t-\sigma\tau}^{0} ds[\exp(-H_0(t-s-\sigma\tau)+Q_0(\tau-\sigma)\ W\ B_{s,\sigma}](a,b,\alpha,\beta). \qquad (2.3.8)$$

Direct differentiation, recalling that $Q = Q_0 + W$, now gives

$$D_{t,\tau}\ B_{t,\tau} = Q\ B_{t,\tau}. \qquad (2.3.9)$$

Clearly $B_{0,0}$ is the identityoperator, and so one may conclude that

$$B_{t,\tau} = \exp(-Ht-Q\tau), \qquad (2.3.10)$$

which establishesthe Feynman-Kac formula (2.3.1).

In terms of $x(s,\sigma)$ and $\Psi(s,\sigma)$ one may write the integral on the right hand side of the Feynman-Kac formula (2.3.1) heuristically as

$$\int DxD\Psi \exp - \int d\sigma \int_0^{t+\sigma\tau} ds\ (\partial x^i/\partial s)\Psi^i + W(x,\Psi)\ . \qquad (2.3.11)$$

The integrand in the exponent in this expression is interesting in its own right as being a supersymmetric analogue of the Lagrangian, generating the square root of the classical equations of motion.

By evaluating covariances one may show that

$$x(s,\sigma) = x(s) + (i\sigma/\sqrt{2})\Psi(s)$$

$$\Psi(s,\sigma) = \Psi(s) - \sqrt{2}\sigma x'(s) \qquad (2.3.12)$$

(where $x(s)$ and $\Psi(s)$ are bosonic and fermionic Brownian bridges).

3. PATH INTEGRALS IN CURVED SUPERSPACE AND THE GAUSS-BONNET-CHERN THEOREM

The Gauss-Bonnet-Chern theorem relates the Euler number of a Riemannian manifold to the integral of its top Euler class. If M is a compact Riemannian manifold of even dimension n=2k, with metric γ, the theorem states that

$$\chi = \int_M E \tag{3.1}$$

where χ is the Euler number of M and ε is a representative of its top Euler class. If dV denotes the volume form on M corresponding to the metric γ one may use

$$\varepsilon = E\, dV \tag{3.2}$$

with the scalar function E given by

$$E = (1/k!)(-1)^k (8\pi)^{-k} \times \sum_{(p,q\ \text{perm}\{1,\dots,n\})} (-1)^p (-1)^q R^{p[1]p[2]}{}_{q[1]q[2]} \cdots R^{p(n-1)p(n)}{}_{q(n-1)q(n)}. \tag{3.3}$$

This theorem was originally proved by Chern in 1944 [13], and is also one of many results contained in the Atiyah-Singer index theorem [14]. Recently very simple, but not completely rigorous, proofs of the Atiyah-Singer index theorem have been given by Alvarez-Gaume [15] and by Friedan and Windey [11], making use of Witten's observations [16] on properties of supersymmetric quantum mechanics. These proofs, which will be referred to as the supersymmetric proofs, have two essential ingredients. First there is McKean and Singers' expression of the Euler number in terms of the supertrace of the operator $\exp(-\Delta t)$ (Δ being the Laplacian on forms on the manifold) [17],

$$\chi = \operatorname{Str} \exp(-\Delta t). \tag{3.4}$$

(This expression was rediscovered by Witten in the language of supersymmetry.) Secondly there is the use of path integral methods (including fermionic path integrals) to calculate this supertrace.

McKean and Singer's formula is perfectly rigorous; the stage at which lack of complete rigour comes in to the supersymmetric proofs is in the path integrals, particularly in the use of fermionic path integrals which are integrals in spaces of paths in some anticommuting space. The purpose of this final lecture is to show how combining the ideas of the first lecture with bosonic curved space path integration one can make the supersymmetric proofs valid in a very straightforward way, so that the clarity and simplicity of the proofs are not lost.

As with other heat kernel proofs of the index theorem (18–24), the strategy of the supersymmetric proofs is to prove a stronger local version of the Gauss-Bonnet-Chern theorem first conjectured by McKean and Singer[17]. This states that

$$\lim_{t\to 0} \operatorname{str}(\exp(-\Delta t))(x,x) = E(x) \tag{3.5}$$

at each point x of M. (Here strdenotes a $2^n x 2^n$ matrix supertrace, whereasStr in equation(1.4) denotesa full supertrace in both the matrix and operator sense.) Standard cutting and pasting arguments mean that one can work in R^n to evaluate the density $\exp(-\Delta t)(x,x)$ in the limit $t\to 0$.

In the first section of this lecture, beginning with the natural correspondence between forms on R^n and functions on (n,n)-dimensional superspace,

$$\sum_{1\leq i_1<\ldots<i_p\leq n} a_{i_1,\ldots,i_p}(x)\, dx^{i_1}\ldots dx^{i_p}$$

$$\leftrightarrow$$

$$\sum_{1\leq i_1<\ldots<i_p\leq n} a_{i_1,\ldots,i_p}(x)\, \theta^{i_1}\ldots\theta^{i_p}, \tag{3.6}$$

it is shown how the Laplacian $\Delta = (d + \delta)^2$ can be expressed as a differential operator on the space of smooth functions on the superspace. This leads to a superspace expression for the density $\operatorname{str}\exp(-\Delta t)(x,x)$. In the next section, the fermionic path integral developed in the first lecture are combined with conventional path integration methods to give a formula for the heat kernel of the superspace version of the Laplacian. The proof of the Gauss-Bonnet-Chern theorem is completed in the last section by using the path integral techniques to evaluate the limit as $t \to 0$ of the matrix supertrace of the superspace heat kernel. The method both picks out the constant term, and shows that the remaining terms tend to zero with t, in a very simple way. More details of this proof may be found in [25].

Methods involving anticommuting variables have mostly been used for formal heuristics, rather than solid analysis. The results of this lecture show that anticommuting variables can be used for rigorous analysis; moreover the process of making the physicist's heuristic approach rigorous does not destroy the beautiful simplicity of the physicist's argument.

Other related proofs of the Gauss-Bonnet-Chern theorem and indeed the full Atiyah-Singer index theorem should be mentioned here, such as Bismut's probabilistic proof [21], and two different analytic proofs, by Getzler [22] and by Cycon,Froese,Kirsch and Simon [23]. There is also the proof by Lott [24], which uses the approach to Fermionic path integration developed by Osterwalder and Schrader [26]. This approach is different from that of these lectures, which takes as literally as possible the idea introduced by physicists of a measure (of Berezin type) in the space of paths in superspace.

3.1 A superspace expression for the Euler number density

As before, let M be an n-dimensional compact Riemannian manifold, with metric g. In this section the density of the Euler number of M will be expressed in terms of the kernel of a differential operator on (n,n) dimensional superspace.

First some general comments about supertraces and kernels are required. There is a natural identification

$$i : L^{2\prime}(B_L^{r,s}) \to (L^2(R^r))^{2^s}$$

with $\quad i(f) = (f_\mu \mid \mu \quad M_s)$

(3.1.2)

if $\quad f(x,\theta) = f_\mu \theta^\mu.$

(3.1.3)

An operator K on $L^{2,}(B_L^{r,s})$ has an integral superkernel if and only if the corresponding operator $i^{-1} \circ K \circ i$ on $(L^2(R^r))^{2^s}$ has a (matrix valued) integral kernel $K(x,y)$. That is, for each x,y in $R^r \times R^r$, $K(x,y)$ is a $2^s \times 2^s$ matrix with complex entries $K_\mu{}^\nu(x,y)$ such that for all x in R^r, (f_μ) in $(L^2(R^r))^{2^s}$

$$\int d^r x \; K_\mu{}^\nu(x,y)\; f_\nu(y) = ((i^{-1} \circ K \circ i)f)_\mu(x).$$

(3.1.4)

The matrix supertrace of $K(x,y)$ is defined by

$$\mathrm{str}\;(K(x,y)) = \sum_{\mu \in M_s} (-1)^{|\mu|}\; K_\mu{}^\mu,$$

(3.1.5)

where $|\mu|$denotes the number of elementsin the multiindex μ.

A key property of the super heat kernel, which may be proved by explicit calculation, is that

$$\int d^n\theta \; K(x,y,\theta,\theta) = \mathrm{str}\; K(x,y).$$

(3.1.6)

The Euler number density of the n-dimensional manifold M with Riemannian metric g will now be expressed in terms of the superkernel of an operator on $H^\infty(B_L^{n,n},C)$. (Cutting and pasting arguments mean that it is not necessary to use any global supermanifold constructions.) There is a natural identification j of $H^\infty(B_L^{n,n},C)$ with $\Omega(R^n)$ (the space of smooth forms on R^n) defined by

$$j(f) = f_\mu dx^\mu$$

(3.1.7)

if $\quad f(x,\theta) = f_\mu \theta^\mu$.

There is a similar identification of $\overline{\Omega}(R^n)$ (the L^2 closure of $\Omega(R^n)$) and $L^{2,}(B_L{}^{n,n},C)$. Clearly an operator K on $\overline{\Omega}(R^n)$ has an integral kernel if and only if the corresponding operator $j^{-1} \circ K \circ j$ on $L^{2,}(B_L{}^{n,n})$ has an integral superkernel. The operator of importance in section 4 is the Laplacian Δ' on $\overline{\Omega}(R^n)$ corresponding to a metric g'. Under the identification j, Δ' is replaced by the operator

$$\begin{aligned}\Delta'' &= j^{-1} \circ \Delta \circ j \\ &= -g^{ij}(\partial_i + \Gamma_{ik}{}^l\theta^k\delta_l)(\partial_j + \Gamma_{jm}{}^n\theta^m\delta_n) \\ &\quad + R^k{}_j\theta^j\delta_k + \\ &\tfrac{1}{2}R^{ij}{}_{kl}\theta^k\theta^l\delta_i\delta_j\end{aligned} \tag{3.1.8}$$

acting on $H^\infty(B_L{}^{r,s},C)$. (Recall that ∂_i denotes differentiation with respect to x^i and δ_i denotes differentiation with respect to θ^i, as in equation (1.3.9)).

Let Δ denote the Laplacian on $\Omega(M)$ corresponding to the metric g. For the proof of the Gauss-Bonnet-Chern theorem given in the next section, it is neccessary to prove the existence of the limit as t tends to zero of str $\exp(-t\Delta)(x,x)$ for each point x of M, and to evaluate this limit. Standard cutting and pasting arguments (described in the book of Cycon, Froese, Kirsch and Simon[23]) show that it is sufficient to work on R^n in the following manner. Suppose that the point x in M lies in a coordinate chart U, with corresponding normal coordinate mapping $\Psi: U \to R^n$ such that $\Psi(x) = 0$, and the closure of $\Psi(U)$ is compact. Then a metric g' on R^n is chosen such that $g' = \Psi^{-1*}(g)$ on $\Psi(U)$ and g' is the Euclidean metric outside a compact set V containing $\Psi(U)$. It can then be shown that, if Δ' is the Laplacian on $\Omega(R^n)$ corresponding to the metric g', then

$$|\exp-\Delta t(x,x)-\exp-\Delta' t(0,0)| = O(\exp-c/t) \tag{3.1.9}$$

where c is a positive constant. Hence, when evaluating χ using equation (1.5) one may replace str $\exp-t\Delta(x,x)$ by str $\exp(-\Delta' t)(0,0)$; combining this with (2.2.7) one obtains

$$\lim_{t\to 0} \text{str}\, \exp-t\Delta(x,x) = \lim_{t\to 0} \int d^n\theta\, \exp(-(\Delta'')t)(0,0,\theta,\theta). \tag{3.1.10}$$

Thus one has an expression for the Euler number density of M in terms of the super kernel of an operator on $L^{2,}(B_L^{n,n})$. In the following sections path integral techniques will be developed to evaluate this super kernel, and hence prove the Gauss-Bonnet-Chern formula.

3.2 Path Integration in curved superspace

In this section a measure on the space of paths in superspace is developed by combining conventional bosonic path integrals with the fermionic path integrals of lecture 1. This measure is then used to derive a Feynman-Kac-Itô formula for the kernel of a wide class of differential operators on $H^\infty(B_L^{r,s},C)$. (In section 3 this formula will be used to prove the Gauss-Bonnet-Chern theorem.)

The measure on the space of paths in the superspace $B_L^{n,2n}$ which will now be described is a generalisation of Wiener measure. This particular measure is designed to make the proof of the Gauss-Bonnet-Chern theorem as straightforward as possible. It will be used in to give a path integral expression for the super kernel of the operator $\exp-\Delta'' t$ (where Δ'' is the localised superspace Laplacian defined in equation (2.3.2)). Similar measures have been used by Elworthy [27] in the purely commuting case.

Definition 3.2.1 (a) Let I be the open interval (0,t), $a,b\in R^n$ and $\alpha,\beta,\kappa\in B_L^{0,n}$. Let g' be a metric on R^n equal to the Euclidean metric outside a compact set V containing the origin 0. Also for $0<s\leq t$ let P(s,x,y) be the kernel of the operator

$\exp -sH_0$ where H_0 is the scalar Laplacian $\partial_i g^{ij}\partial_j$ acting on $C^\infty(R^n)$. The super measure $d\mu[a,b,\alpha,\beta,\kappa,t]$ is defined by the finite distribution functions $f_J \in H^\infty(B_L^{(n,2n)J},C_L)$ where, if $J \subset I$ with $J=\{t_1,\ldots,t_N\}$ and $0<t^1<\ldots<t^N<t$,

$$\begin{aligned} & f_J(x^{1i},\theta^{1i},\rho_{1i},\ldots,x^{Ni},\theta^{Ni},\rho_{Ni}) \\ &= P(t_1,a,x^1)P(t_2-t_1,x^1,x^2)\ldots P(t-t_N,\ x^N,b) \\ &\quad \times \exp -i\{(\kappa_i(\theta^{1i}-\alpha^i)+\rho_{i1}(\theta^{2i}-\theta^{1i})+\ldots+\rho_{Ni}(\beta^i-\theta^{Ni})\} \end{aligned} \tag{3.2.1}$$

(It may easily be checked, using the semigroup property of $\exp -tH_0$ that the finite distributions f_J satisfy the consistency conditions (i) and (ii) of definition 3.1.1, and thus define a super measure.)

Using this measure, a key theorem, which establishes a Feynman-Kac-Ito formula, that is, a path integral formula for the heat kernel of certain differential operators on the space of L^{2}' functions on the superspace $B_L^{n,n}$ can be proved. This formula will be used in the next section to prove the Gauss-Bonnet-Chern theorem. Here the theorem will simply be stated, full details of the proof may be found in [25].

<u>Theorem 3.2.2</u> For $i=1,\ldots,n$ and $\mu,\nu \in M_n$ let $h_{i\mu}{}^{\nu}$ and $K_\mu{}^\nu$ be complex valued C^∞ functions on R^n with compact support. Let g be a Riemannian metric on R^n equal to the Euclidean metric outside a compact region. Then, if H is the closure of the differential operator on $C_0^{\infty}{}'(B_L^{n,n},C_L)$ defined by

$$H = \partial_i g^{ij}\partial_j + h_{i\mu}{}^{\nu}\theta^\mu\delta_\nu\partial_i + K_\mu{}^\rho\theta^\mu\delta_\nu, \tag{3.2.2}$$

$$\exp(-Ht)(a,b,\alpha,\beta)$$
$$= \int d^n\kappa \int d\mu[a,b,\alpha,\beta,\kappa]$$
$$\times \{\exp- \int_0^t h_{i\mu}{}^{\nu}(x(s))\theta^{\mu}(s)\rho_{\nu}(s)g_{ij}(x(s))\, dx^j(s)$$
$$+ \int_0^t (K_{\mu}{}^{\nu}(x(s))-\partial_i h^i{}_{\mu\nu}(x(s))\theta^{\mu}(s)\rho_{\nu}(s)\}\, ds. \tag{3.2.3}$$

Here the random variable

$$\exp(\int_0^t h_{i\mu}{}^{\nu}(x(u))\theta^{\mu}(u)\rho_{\nu}(u)dx^i(u) + \int_0^t k_{\mu}{}^{\nu}(x(u))\theta^{\mu}(u)\rho_{\nu}(u)\, du)$$

in the integrand is defined as follows. For each N=1,2,... let

$$J(N) = \{rt/2^N \mid r=1,\dots,2^{N-1}\} \tag{3.2.4}$$

and let functions g_N in $C_0^{\infty}{}'(B_L{}^{(n,2n)J(N)},C_L)$ be defined by

$$g_N(x^{1i},\theta^{1i},\rho_{1i},\dots,x^{Ni},\theta^{Ni},\rho_{Ni})$$

$$= \exp\{\sum_{r=0}^{2^{N-1}} [\tfrac{1}{2}(h_{i\mu}{}^{\nu}(x^r)\theta^{r\mu}\rho_{r\nu}+h_{i\mu}{}^{\nu}(x^{r+1})\theta^{r\mu}\rho_{r\nu})(x^{ri}-x^{r+1,i})$$
$$+ (t/2^N)\, K_{\mu}{}^{\nu}(x^r)\, \theta^{r\mu}\, \rho_{r\nu}]\} \tag{3.2.5}$$

(where, for notational convenience, $x_0=a$, $x_2 = b$, $\theta_0=\alpha$, $\theta_2 = \beta$ and $\rho_0=\kappa$.) Then the sequence $g_1, g_2, \dots$ defines a super random variable on the super probability space $(B_L{}^{(n,2n)I},\{f_J\},d\mu)$.

It is also proved in [25] that this does indeed define a random variable according to definition 2. In the final part of this lecture the Feynman-Kac-Ito formula will be used to prove the Gauss-Bonnet-Chern theorem.

3.3. Proving the Gauss-Bonnet-Chern theorem

In this section the local Gauss-Bonnet-Chern theorem is proved by using the superspace Feynman-Kac-Ito formula (Theorem 3.2.2) to evaluate the super kernel of $\exp(-\Delta''t)$, Δ'' being the superspace version of the localised Laplacian.

Theorem 3.3.1

Let M be a compact Riemannian manifold, with metric g and even dimension n=2k, and let Δ denote the Laplacian on forms on M. Then at each point x of M

$$\lim_{t\to 0} \operatorname{str} \exp(-\Delta t)(x,x) = E(x) \tag{3.3.1}$$

where, as in equation (3.3), E is the scalar function

$$E = (1/k!)(-1)^k(8\pi)^{-k} \times \sum_{p,q\in \mathrm{perm}\{1,\ldots,n\}} (-1)^p(-1)^q\, R^{p[1]p[2]}{}_{q[1]q[2]}\cdots R^{p(n-1)p(n)}{}_{q(n-1)q(n)}. \tag{3.3.2}$$

(The Gauss-Bonnet-Chern theorem (equation (3.1)) follows immediately from (3.3.1) together with McKean and Singer's formula (3.4)[17]. (3.3.1) is a stronger, local result originally conjectured by McKean and Singer, and first proved by Patodi[18].)

Proof (The idea of the proof is to rescale the odd parameters which occur in the superspace calculation of the supertrace, and to use the fact that $t^{n/2}\,P(t,0,0)$ remains finite as t tends to zero, to isolate the contribution to the supertrace which survives in this limit from the contribution which dies away; the use of the heat kernel of the scalar Laplacian, rather than the more geometric Bochner Laaplacian, makes this separation very simple.)

With the notation of section 2.3,

$$\operatorname{str} \exp(-\Delta t)(x,x) = \int d^n\alpha\, \exp(-\Delta''t)(0,0,\alpha,\alpha). \tag{3.3.3}$$

Now Δ'' has the form

$$\Delta'' = -\partial_i g^{ij}\partial_j + S^i{}_k{}^l\theta^k\delta_l\partial_i + T_k{}^l\theta^k\delta_l$$

$$+ U_{kl}{}^{mn}\theta^k\theta^l\delta_m\delta_n,$$

(3.3.4)

where $U_{kl}{}^{mn} = \frac{1}{2}R'{}_{kl}{}^{mn} - g_{ij}\Gamma'{}_{ik}{}^{m}\Gamma'{}_{j}{}^{ln}$, and $S^i{}_k{}^l$, $T_k{}^l$ (and of course $U_{kl}{}^{mn}$) are smooth compact support functions on R^n. ($\Gamma'{}_{ij}{}^k$ and $R'{}_{ij}{}^{kl}$ are the connection and curvature coefficients corresponding to the metric g'. The precise form of $S^i{}_k{}^l$ and $T_k{}^l$ in terms of curvature and connection coefficients will not be important.) Thus, using Theorem 3.3.1, with $H = \Delta''$,

$$\text{str}\exp(-tD)(x,x) = \int d^n\kappa\, d^n\alpha\, d\mu[0,0,\alpha,\alpha,\kappa,t]$$
$$\times \exp -\{\int_0^t S_{ik}{}^l(x(s))\theta^k(s)\rho_l(s)\,dx^i$$
$$+ \int_0^t [T_k{}^l(x(s))\theta^k(s)\rho_l(s) + U_{kl}{}^{mn}(x(s))\theta^k(s)\theta^l(s)\rho_m(s)\rho_n(s)]ds\}.$$

(3.3.5)

Inspection of the fermionic part of the measures $d\mu[0,0,\alpha,\alpha,\kappa,t]$ and $d\mu[0,0,0,0,0,t]$ shows that this equation may be rewritten as

$$\text{str}\exp(-\Delta t)(x,x) = \int d^n\kappa\, d^n\alpha\, d\mu[0,0,0,0,0,t]$$
$$\times \exp -\{\int_0^t S_{ik}{}^l(x(s))(\theta^k(s)+\alpha^k)(\rho_l(s)+\kappa_l)\,dx^i$$
$$+ \int_0^t [T_k{}^l(x(s))(\theta^k(s)+\alpha^k)(\rho_l(s)+\kappa_l)$$
$$+ U_{kl}{}^{mn}(x(s))(\theta^k(s)+\alpha^k)(\theta^l(s)+\alpha^l)(\rho_m(s)+\kappa_m)(\rho_n(s)+\kappa_n)]ds\}.$$

(3.3.6)

Setting $\alpha = t^{-1/4}\alpha'$, $\kappa = t^{-1/4}\kappa'$ (and hence $d\alpha = t^{1/4}\,d\alpha'$ and $d\kappa = t^{1/4}\,d\kappa'$) gives

$$\text{str}\exp(-\Delta t)(x,x) = \int d^n\kappa'\, d^n\alpha'\, t^{n/2}\, d\mu[0,0,0,0,0,t]$$
$$\times \exp-\{\int_0^t t^{-1}\, U_{kl}{}^{mn}[0]\alpha^k\alpha^l\kappa_m\kappa_n\, ds\},$$

(3.3.7)

where the f_{ai} and g_a are of compact support, f_{4i} is zero, and (for a=1,2,3) $f_{ai}(0,\theta,\rho)=0$ for all (θ,ρ) in $B_L^{0,2n}$; and also, for a = 1,2,3,4, $g_a(0,\theta,\rho)$ is zero for all (θ,ρ) in $B_L^{0,2n}$. (These properties of g_{ai} and f_a follow from the fact that normal coordinates are being used on M at x.)

Now it is shown in [25] that the terms

$$\sum_{a=0}^{4}\Big[\int_0^t t^{-a/4} f_{ai}(x(s),\theta(s),\rho(s))\, dx^i + \int_0^t t^{-a/4} g_a(x(s),\theta(s),\rho(s))\Big]ds\},$$

make no contribution in the limit as t tends to zero. Thus

$$\lim_{t\to 0} \operatorname{str} \exp(-\Delta t)(x,x)$$

$$= \lim_{t\to 0} \int d^n\kappa\, d^n\alpha\, d\mu[0,0,0,0,0,t] \exp -U_{kl}{}^{mn}(0)\alpha^k\alpha^l\kappa_m\kappa_n$$

$$=(4\pi)^{-k} \int d^n\kappa\, d^n\alpha \exp -\tfrac{1}{2}R_{kl}{}^{mn}(0)\alpha^k\alpha^l\kappa_m\kappa^n$$

$$= (1/k!)(-1)^k(8\pi)^{-k}$$

$$\times \sum_{p,q\in \operatorname{perm}\{1,\dots,n\}} (-1)^p(-1)^q\, R^{p[1]p[2]}{}_{q[1]q[2]}(x)\dots R^{p(n-1)p(n)}{}_{q(n-1)q(n)}(x), \tag{3.3.8}$$

as required. ∎

3.4 Concluding remarks

These lectures describe a new approach to path integration for theories involving fermions. The aim throughout has been to make the heuristic approach used by physicists, based on a fairly concrete picture of paths in superspace, rigorous in as direct a manner as possible; this does seem to lead to some genuinely new analytic techniques. An obvious area for future development is the extension of the work from quantum mechanics to field theory and string theory.

Some other work on fermionic path integrals, in addition to that already cited, is listed in reference 28. Inevitably, this is not a complete list.

Acknowledgement

The author is greatful to the Science and Engineering Reasearch Council of Great Britain for support under advanced research fellowship B/AF/687.

REFERENCES

1. Simon,B.: Functional integration and quantum mechanics. New York,San Francisco and London:Academic Press 1979.

2. Bochner,S.: Stochastic Processes, Ann of Math 48, 1014-1061[1947].

3. Dashen,R.F.,Hasslacher,B. and Neveu,A.: Semiclassical bound states in an asympyotically free theory, Phys.Rev.D12,2443-2458, 1975.

4. It is not possible to give a full list; some examples are:

Applebaum,D. and Hudson,R.L.:Fermion Ito's formula and stochastic evolutions, Comm.Math.Phys. 96,473[1984].

Barnett,C.,Streater,R.F. and Wilde,I.F.:The Ito-Clifford integral, Journ.Func.Anal.48,172-212[1982].

Hudson,R. and Lindsay,J.M.:A non-commutative martingale representation theorem for non-Fock quantum Brownian motion, Journ.Func.Anal.61,202-221[1985].

Kümmerer,B.,Markov dilations on W *-algebras, Journ. Func.Anal.63,139[1985].

See also: Quantum probability and its applications, eds. Accardi,L. and von Waldenfels,W., Lecture Notes in Mathematics 1136.

5. Haba,Z.: Euclidean formulation of quantum spinning particles, Bielefeld preprint no 214,[1986].

6. Martin,J.L.: The Feynman principle for a Fermi system, Proc.Royal Soc. A251,543-549,1959.

7. Berezin,F.A. and Marinov,M.S.:ZhETFPis.Red.21,678-180[1975] =JETP letters 21,320-321[1975].

Brink,L., Deser,S., Zumino,B., Di Vecchia,P., and Howe,P.: Physics Letters 64B[1976]435.

Di Vecchia,P.,Brink,L., and Howe,P.: Nuclear Physics 118B [1976]77.

8. Rogers,A.: Graded manifolds, supermanifolds and infinite dimensional Grassmann algebras, Comm.Math.Phys.105,375-364[1986].

9. Berezin,F.A.: The method of second quantization, New York:Academic Press 1966.

10. Rogers,A.: Fermionic path integration and Grassmann Brownian motion, Comm.Math.Phys. 113,353-368[1987].

11. Friedan,D. and Windey,P.: Supersymmetric derivation of the Atiyah-Singer index theorem and the chiral anomaly, Nuc.Phys.B235,395-416[1984].

12. Guerra,F. Rosen,L. and Simon,B.: The $P(\phi)_2$ Euclidean Quantum Field Theory, Ann. of Math. 101[1975],111-259.

13. Chern, S.S.: A simple intrinsic proof of the Gauss-Bonnet formula for closed Riemannian manifolds. Ann.Math.45,747-752[1944].

14. Atiyah, M.F. and Singer, I.M., The index of elliptic operators on compact manifolds. Bull.Am.Math.Soc.69,422-433[1963].

15. Alvarez-Gaume, L.: Supersymmetry and the Atiyah-Singer index theorem. Comm.Math.Phys.90,161-173,[1983].

16. Witten,E.: Constraints on supersymmetry breaking, Nucl.Phys.B202, 253[1982].

17. McKean,H. and Singer,I.M.: Curvature and eigenvalues of the Laplacian. J.Diff.Geo.1,43-69[1967].

18. Patodi,V.K.: Curvature and eigenforms of the Laplace operator. J.Diff.Geo.5,251-283[1971].

19. Gilkey,p.: Curvature and the eigenvalues of the Laplacian, Adv.Math.10,344-382[1973].

20. Atiyah,M.F., Bott,R. and Patodi,V.K.: On the heat equation and the index theorem. Inv.Math.19,279-330[1973].

21. Bismut,J.M.: The Atiyah-Singer theorems: a probabilistic approach, I. The index theorem, J.Funct.Anal.57,56-99[1984].

22. Getzler,E.: A short proof of the Atiyah-Singer index theorems. Topology 25,111-117[1986].

23. Cycon,H.L., Froese,R.G., Kirsch,W. and Simon,B.: Schrodinger operators with applications to quantum mechanics and global geometry. Springer Verlag, Berlin Heidelberg[1987].

24. Lott,J.: Supersymmetric Path Integrals, Comm.Math.Phys.100,605-629[1987].

25. Rogers,A.: A superspace path integral proof of the Gauss-Bonnet-Chern theorem, King's College preprint 1987.

26. Osterwalder,K. and Schrader,N.: Euclidean fermi fields and a Feynman-Kac formula for Boson-Fermion models. Helv.Phys.Acta46, 277[1973].

27. Elworthy,D.K.: Path integration on manifolds. In: Mathematical Aspects of Supergravity: eds H.-J. Seifert, C.J.S.Clarke and A.Rosenblum, Reidel, Dordrecht, [1984].

28. Seiler,E.: Constructive quantum field theory:fermions, in:Gauge theories:Fundamental Interactions and rigorous results, eds. Dita,P., Georgescu,V. and Purice,R., Boston,Basel,Stuttgart:Birkhauser[1982](and refernces therein).

Kupsch,J. Measures for fermionic integration, Universitat Kaiserslautern preprint, April 1986.

Segal,I.E.: A non-commutative extension of abstract integration, Ann. of Math.58,595-596,[1953].

Matthews,P.T.and Abdus Salam:propogators of quantised fields, Nuovo Cimento 2,120-134[1955].

Gross,L.: On the formula of Mathews and Salam, J.Func.Anal.25, 162-209[1977].

Lecture for the 7th Scheveningen conference on "The mathematical structure of field theories" : August 24th, 1987.

SUPERMANIFOLDS, SUPERSYMMETRY AND BEREZIN INTEGRATION.

Peter Bryant

Department of Pure Mathematics and Mathematical Statistics,

Cambridge University, 16 Mill Lane,

Cambridge, United Kingdom CB1 1SB.

Introduction

Supermanifold theory has traditionally developed in an uneasy alliance with what physicists call supersymmetry. Yet it is only comparatively recently ([F] 1986 , [dW] 1984 , [S] 1984) that the superspace of Salam & Strathdee [SS] has been viewed as a supermanifold by physicists. Since then the emergence of new mathematical techniques (albeit for a seemingly limited class of supermanifolds) has made the goal of modelling supersymmetric field theories that much more accessible.

In this lecture I draw together some more or less familiar results on supermanifolds and superspace ending with a brief account of Berezin integration (from a geometrical standpoint). This may be taken as an attempt to supplement Freund's book (particularly chapters 6,7,8 and 9) or as an introduction to the subject. I won't have time to really get down to tackling the various interpretations of field theories of current interest but I shall end with a (very subjective) list of areas of physical interest in which supermanifolds have already made some impact and in which future developments are likely.

There are perhaps two principles to bear in mind in the discussion. The first is that the very existence of (any) supermanifold theory means that ***classical field theories with anticommuting variables can and do make sense.*** This is hardly controversial since the Cartan calculus for (vector valued) differential forms is a familiar and useful tool in gauge theory. Indeed supermanifold theory is sometimes viewed as little more than a way of unifying the Cartan calculus on some underlying manifold with the set of spin structures on it : see e.g. [Bu]. Then the ground ring is merely a bookkeeping device and taking components (or going to the body) reveals the true geometry.

The second principle is that *even and odd variables should be placed on the same footing*. This leads to the study of supermanifolds which are more than merely formal extensions of (vector bundles over) their bodies. Although the global topology of such (non-de Witt) supermanifolds is restricted by a rich foliation structure [RC], they are cohomologically nontrivial when we consider cohomology with vales in their structure sheaves [Bz]. Further a body may not even exist in the most extreme examples.

This second principle has a striking corollary when we view (4,4)-superspace as a supermanifold : the (real-valued) coordinates x^i on Minkowski space M_0 need to be reinterpreted as "c-number"-valued coordinates (the use of the phrase "c-number" in this context disguises this need). As we indicate next, this step is forced if inconsistencies are to be avoided.

The **super Poincare group paradox** may be described as follows :-

If $\{Q,\bar{Q}\} = T$ implies $[\varepsilon_1 Q, \varepsilon_2 \bar{Q}] = \varepsilon_1 \varepsilon_2 T$ and $(\varepsilon_1 \varepsilon_2)^2 = -(\varepsilon_1)^2 (\varepsilon_2)^2 = 0$ then in what sense is it legitimate to say that "the commutator of two supersymmetries is a spacetime translation" ??

Here of course T is a translation generator (and I suppress indices and normalisation constants for clarity), ε_1 and ε_2 are constant odd parameters. The traditional reason for changing the anticommutators to commutators is that one knows how to exponentiate Lie brackets and it is the introduction of the odd paramaters that causes the difficulties.

There are two ways to resolve this paradox (*both* of which involve the reinterpretation of spacetime coordinates mentioned above). Firstly one can learn how to "exponentiate graded Lie algebras" (or rather find a differential operator representation of graded Lie algebras) directly without introduction of an auxiliary space of odd (and then also even) parameters. This is precisely what Kostant's category of graded manifolds [K] is designed to do.

The second way (and the approach we shall be following) is to give up the idea that the finite transformations associated to the commutator bracket are ordinary translations. By introducing an auxiliary set of odd and even parameters (ground ring) one automatically enlarges the group of automorphisms. One says that $\varepsilon_1 \varepsilon_2 T$ is the generator of a supertranslation whose parameter $\varepsilon_1 \varepsilon_2$ is pure soul. This approach is the supermanifold approach of Rogers [Ro1] and others.

Basic supermanifold theory.

The first step in specifying a supermanifold theory is to *choose a ground ring* (this is at least a topological algebra B equipped with an algebra map $\varepsilon_0 : B \to R$ with B assumed to be local in the sense that $b \in B$ is invertible iff $\varepsilon_0(b)$ is nonzero). This is the set of "commuting and anticommuting numbers" and is usually taken to be a Grassmann algebra.

For technical reasons I will only consider finite Grassmann algebras here (consequently I consider only classical theories) : $B_L = \Lambda R^L$ together with the fine topology - i.e. its topology as a vector space - [Ro1] (where L is assumed to be at least twice as large as the number of arithmetic operations to be performed). The point is that L should have no physical significance and one anticipates that preparation for quantisation will include taking the formal limit $L \to \infty$ [Ro5].

Supermanifold theories over infinite dimensional Grassmann (and other) algebras have been developed [dW,JP,Ro1] but then we lose one of our strongest weapons which enabled us to control the superstructure for supermanifolds over B_L. This is due to the fact that such supermanifolds are automatically ordinary manifolds of very high dimension and thus ordinary manifold techniques apply (e.g. taking universal covers, foliation structure etc : see [RC]).

Alternatively one may formally delete dependence on the ground ring by considering supermanifolds as *functors* from certain ground rings to topological spaces (with extra structure) [Sch2]. As an example every graded manifold can be considered as a functor from finite dimensional Grassmann algebras to Rogers supermanifolds [Ba1]. When pressed we take a formal limit [as in Ro5] obtaining a theory similar to [dW] but where the formal B_∞ has a Frechet topology [MK] ; details in [Bry2]. This is attractive from an algebraic point of view but we lose sight of the kind of supersmoothness that comes automatically with Banach ground rings.

Taking B as the ground ring write $B^{p,q}=B_0\times\cdots\times B_0\times B_1\times\ldots\times B_1$ for the associated *superEuclidean space* (so B and $B^{1,1}$ are the same thing). These are the analogues of (and play a similar role to) the Euclidean spaces R^k in ordinary manifold theory.

Definition

A supermanifold of dimension (p,q) is a (hausdorff , paracompact) topological space locally modelled on $B^{p,q}$ with "supersmooth transition functions".

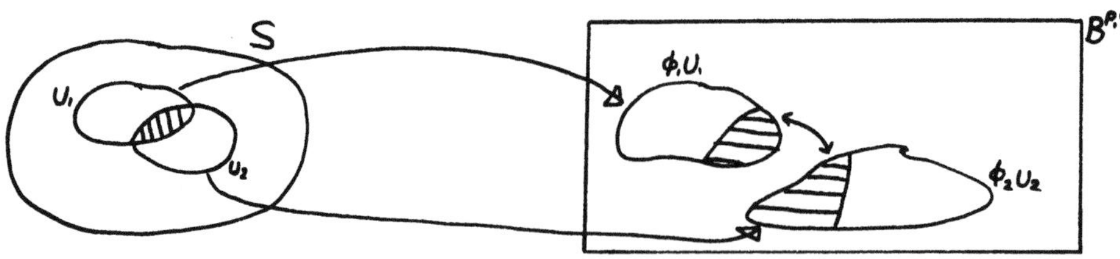

supercharts on S

$\phi_{ij}:\tau_i(U_i\cap U_j)\to\tau_j(U_i\cap U_j)$ assumed to be supersmooth in some appropriate sense (typically they are smooth with B_0-linear differentials). This is the subject of our next definition.

Most people use the G^∞-supermanifolds of Rogers [Ro1] although the full theory needs careful handling (sheaves of derivations - analogues of vector fields - are not locally free).

Use notation [K] B_L is spanned by a set $\{e_{(\mu)}\}$ indexed by

$$M_L = \{\emptyset \text{ or } (\mu)=(\mu_1,\mu_2,\cdots\mu_k):1\le\mu_1<\mu_2<\cdots<\mu_k\le L, \text{k variable integer less than L}\}.$$

Let $\theta^1.....\theta^q$: $B^{p,q}\to B$ (the odd projections) be the odd coordinates for the supermanifold $B^{p,q}$. Smooth functions are assumed to be suitably analytic in these odd directions.

Definition

F:$B^{p,q}\to B$ is said to be G^∞ if we have

$$F(x,\theta)=f_0(x)+f_\alpha(x)\theta^\alpha+\cdots f_{top}(x)\theta^1.\theta^2.....\theta^q \qquad \text{(z-expansion)}$$

with each coefficient f (a function on the even variables alone) completely determined by its restriction $\tilde{f}$ to $R^p\subset B^{p,q}$ ($\tilde{f}\in C^\infty(R^p,B_L)$) through the Taylor expansion

$$f(x)=\tilde{f}(x_0)+(x^i-x_0^i)\partial_i\tilde{f}(x_0)+\cdots = Z\tilde{f}(x).$$

(This is a finite sum because $x-x_0$ is nilpotent).

Remark: It is often more convenient to use the classes GH^∞ or H^∞ defined in [Ro5] rather than G^∞. For H^∞ the only change in the definition is that we require that each coefficient in the z-expansion be generated by a *real valued* smooth function. Compare with [Ba1].

Example (H^∞-superfields in $B^{2,2}$)

A superfield F:$B^{2,2}\to B$ is determined by four real valued fields $\phi,\psi_1,\psi_2,A\in C^\infty(R^2)$

$$F(x;\theta)=F(x,y;\theta^1,\theta^2)=Z\phi(x,y)+Z\psi_1(x,y)\theta^1+Z\psi_2(x,y)\theta^2+ZA(x,y)\theta^1\theta^2$$

$$\text{(e.g.) } Z\phi(x,y)=\phi(x_0,y_0)+(x-x_0)\frac{\partial\phi}{\partial x}(x_0,y_0)+(y-y_0)\frac{\partial\phi}{\partial y}(x_0,y_0)+1/2(x-x_0)(y-y_0)\frac{\partial^2\phi}{\partial x\partial y}(x_0,y_0).$$

Physicists superfields usually have additional Lorentz properties [F] with "even" components (ϕ , A) being even and "odd" components (ψ_1,ψ_2) odd (so overall the superfield is purely even). This can easily be achieved using GH^∞ or G^∞ rather than H^∞ (Lorentz properties are considered a little later). There are two points to bear in mind about this approach

I) In the superfield expansion $\Phi(x,\theta)=\phi(x)+\psi_1\theta^1+\psi_1\theta^2+A(x)\theta^1\theta^2$ the ordinary variables x need to be extended to even Grassmann valued variables.

II) The Taylor series expansion $\phi\to Z\,\phi$ allows us to pass from R^p to $B^{p,0}$ and promote the superfield expansion to an expression defining a genuine supersmooth function on $B^{p,q}$.

In the H^∞ picture the two expansions convey precisely the same information whereas in the G^∞ picture there is some redundancy (the map $Z: C^\infty(R^p) \otimes \Lambda R^q \otimes B_L \to G^\infty(B^{p,q})$ is not injective).

Kostant's theory of graded manifolds [K] extends functions (and coordinates) on X without extending the points of the base space X. Odd coordinates θ^α enter in the form of local generators for a sheaf of graded-commutative algebras over X (actually as local generating sections of an exterior bundle over X). Recall that R^p sits inside $B^{p,q}$ and $C^\infty(R^p) \otimes \Lambda R^q$ is in one-one correspondence with $H^\infty(B^{p,q})$ via the Taylor series. Graded manifolds have been shown to be (globally) equivalent to H^∞-de Witt supermanifolds ([Ba1],[Ro1],[Bry2]), every abstract graded manifold sheaf being reinterpreted as a sheaf of H^∞ functions. It thus makes sense to talk of supersymmetries directly in the category of graded manifolds (see [DS]) except that the anticommutative nature of the parameters of supersymmetry transformations is inadequately represented (supersymmetries are not H^∞).

Graded manifolds and the de Witt condition - charts are coarse open i.e. cylinders - are important in supermanifold theory because the correspondence between them provides a wealth of examples of supermanifolds all of which are attractive both topologically and algebraically. de Witt supermanifolds are noncompact (whenever L>1 or q>0) It is much harder to find compact supermanifolds - one example is the torus with several distinct superstructures, another is the Mobius band. This partly explains the inadequacy of our knowledge about non-de Witt supermanifolds.

We give a geometrical form of the extension of graded manifolds to supermanifolds [Ba3,Bry2]. Let X be an ordinary manifold and choose any (smooth) vector bundle E over X. Then $(X, \Gamma_X(-, \Lambda E))$ turns out to be the most general graded manifold [Ba 2] (up to isomorphism). First extend the bosonic coordinates of X by extending X (of dimension p) to a supermanifold Z(X) of dimension (p,0). Z(X) is the only de Witt supermanifold with body X which has this dimension. Of course we take $Z(R^p) = B^{p,0}$. Z(X) is also characterised by the fact that Taylor's expansion gives an isomorphism $C^\infty(X) = H^\infty(Z(X))$ [Bry2]. By remarks above the exact details of the construction of Z(X) aren't important if we are only interested in the correspondence between superfield expansions.

The supermanifold S = $z(X, \Lambda E)$ [corresponding to the graded manifold $(X, \Gamma_X(-, \Lambda E))$] is the total space of the (vector bundle) pull-back diagram

$$\begin{array}{ccc} S & \to & E^* \otimes B_1 \\ \downarrow & & \downarrow \\ Z(X) & \to & X \end{array}$$

S is a vector bundle over its body manifold X which is in fact characteristic of H^∞ de Witt supermanifolds. All H^∞-de

Witt supermanifolds arise this way.

Remark.

I have recently shown that all (supersmooth) de Witt supermanifolds are H^∞ (and therefore given by the above construction) [Bry4]. This implies the far from trivial remark that every de Witt supermanifold has the global topology of a vector bundle over its body. [More precisely it is not difficult to see that de Witt supermanifolds are fibre bundles over their bodies with vector space fibres. The result mentioned shows that there *does* exist a superatlas whose transition functions act linearly on all fibres].

Superspace.

We describe (4,4)-superspace carefully.

All (H^∞-de Witt) super Lie groups are described as follows [Ro2,K]. Begin with a (Z_2)-graded Lie algebra $\mathfrak{g}=\mathfrak{g}_0+\mathfrak{g}_1$ (i.e. the graded bracket [,] turns $\mathfrak{g}_0$ into an ordinary Lie algebra and $\mathfrak{g}_1$ into a $\mathfrak{g}_0$-module with a symmetric form [,]:$\mathfrak{g}_1\times\mathfrak{g}_1 \rightarrow \mathfrak{g}_0$. The graded Jacobi identity relates these structures).

Construct the ordinary Lie algebra (of high dimension) $\mathfrak{g}B=(\mathfrak{g}\otimes B)_0=\mathfrak{g}_0\otimes B_0+\mathfrak{g}_1\otimes B_1$. Then the associated (connected and simply connected) Lie group G has a natural supermanifold structure which makes it a super Lie group. Indeed if dim $\mathfrak{g}$ = (p,q) then dim $\mathfrak{g}B = 2^{(L-1)}(p+q)$ = dim G and the vector space $\mathfrak{g}B$ is isomorphic to $B_L^{p,q}$. Elements of G take the form $\exp(\lambda P+\mu Q)$ suppressing summation over even generators (i.e. indices on P) and odd generators (i.e. indices on Q). Each λ is a commuting number $\lambda\in B_0$ and each μ is an anticommuting number $\mu\in B_1$.

The body manifold of G is the ordinary Lie group G_0 (that associated to the Lie algebra $\mathfrak{g}_0$ in fact) and G is the (H^∞-de Witt) supermanifold associated to the graded Lie group [K] determined by G_0 and $\mathfrak{g}$ [Ro2]. Then G is constructed (together with its superstructure) by the method above where the vector bundle E over G_0 is the *trivial* vector bundle with fibre $\mathfrak{g}_1^*$ [Bry2].

Then we have the following trivial pull-back diagram

$$\begin{array}{ccc} G=Z(G_0)\times(\mathfrak{g}_1\otimes B_1) & \rightarrow & G_0\times(\mathfrak{g}_1\otimes B_1) \\ \downarrow & & \downarrow \\ Z(G_0) & \rightarrow & G_0 \end{array}$$

A typical element of G_0 is of the form $\exp(\lambda P)$ with $\lambda\in$ R, a typical element of $Z(G_0)$ is of the form $\exp(\lambda P)$ with $\lambda\in B_0$, and a typical element of G is $\exp(\lambda P+\mu Q)$ where in addition $\mu\in B_1$. This is consistent with the identification G = $Z(G_0)\times(\mathfrak{g}\otimes B_1)$ because the expansion of the exponential of an odd operator stops at the first term.

Graded quotient/homogeneous spaces are constructed from pairs $(H_0, \mathfrak{h})$ consisting of a Lie subgroup H_0 of G_0 and a graded Lie subalgebra $\mathfrak{h}$ of $\mathfrak{g}$ with $Lie(H_0)=\mathfrak{h}_0$. Superquotient (or homogeneous) spaces are constructed as the ordinary quotients spaces of the corresponding super Lie groups G and H say. G/H is a supermanifold with body G_0/H_0. Remark: since many of the superspaces used in the physics literature are actually quotient spaces we see that the de Witt supermanifolds form a class of supermanifolds "of physical interest". To obtain others we may take the quotient of a supermanifold under the (properly discontinuous and supersmooth) action of an (ordinary discrete) group [BBry][RC2]. Apart from the class of (de Witt) supermanifolds defined by the vanishing of some superpolynomial (i.e. superalgebraic varieties) these quotient spaces are almost the only known examples of supermanifolds [RC1]. A concrete example is the torus with its standard G^∞-structure [Ro1].

Superspace is the quotient of the super Poincare group by the Lorentz group. It is clear from the above remarks that many of its properties can be deduced from an analysis of the super Poincare algebra.

$$spa = (\ell + t) + (q + \bar{q})$$

where ℓ is the Lorentz algebra, t is the translation algebra and $q, \bar{q}$ are spinorial generators of the odd sector.

Bracket relations are summarised in

$i)$ $[\ell,\ell] \subset \ell$; $[\ell,q] \subset q$; $[\ell,\bar{q}] \subset \bar{q}$ (infinitesimal spinor representation).

ii) $\{q,\bar{q}\} \subset t$.

iii) $[\,*\,,t] = 0$; $\{q,q\} = 0$; $\{\bar{q},\bar{q}\} = 0$.

and spa is the semidirect sum $spa = \ell \lozenge (Ct + q + \bar{q})$.

Write S for **Superspace** S = SPG/L (here Lie(L) = $\ell \otimes B_0$ rather than ℓ)

Since ℓ is not an ideal in spa, L is not a normal subgroup of SPG - if it was L could not have a residual action on superspace. However, as a subgroup of SPG, S has a *nonabelian* group structure [associated with the graded Lie algebra $t+(q+\bar{q})$ whose single non-trivial bracket $\{q,\bar{q}\} \subset t$]. Of course, the projection SPG $\to$ SPG/L = S is not a group homomorphism.

Explicitly the super quotient space is

$$S = Z(M_0) \times F$$

where M_0 is ordinary Minkowski space and F is the vectorspace F = $(q+\bar{q}) \otimes B_1$. Choosing coordinates in Minkowski space ($M_0 = R^4$) and a basis $\{Q_1, Q_2, \bar{Q}_1, \bar{Q}_2\}$ in $q+\bar{q}$ produces the identification $S = B^{4,4}$.

Odd coordinates $\theta^1, \theta^2, \bar{\theta}^1, \bar{\theta}^2 : S \to B$ are chosen so that the odd (or horizontal) projection

$h: S = Z(M_0) \times F \to F$ is given by

$h(s) = \theta^1(s)Q_1 + \theta^2(s)Q_2 + \bar{\theta}^1(s)\bar{Q}_1 + \bar{\theta}^2(s)\bar{Q}_2$. (*)

Since these odd coordinates θ are B_1-valued fields (i.e. odd superfields), there is an induced Lorentz action [namely $\Lambda.\theta(s) = \theta(\Lambda^{-1}.s)$] on them. (*) implies that the odd coordinates carry Weyl chiral and antichiral spinor representations of the Lorentz group and this is the origin of the phrase "odd spinorial coordinates".

The horizontal projection $h: S \to F$ is of interest for a second reason. It is not hard to see that *h is a principal Tr-bundle* (Tr is the translation group associated to t). Further, the splitting $t+(q+\bar{q})$ is reductive, $[t, q+\bar{q}] \subset q+\bar{q}$, and we have a canonical method for writing down a Tr-invariant connection μ^F on the principal bundle $h: S \to F$. One takes the Maurer-Cartan form μ on (the super Lie group) S and uses the projection of $t \otimes B_0 + (q+\bar{q}) \otimes B_1$ onto $(q+\bar{q}) \otimes B_1 = F$. μ is the (unique) left invariant $(t \otimes B_0 + F)$-valued one form on S (a manifestation of the difference between left and right multiplication on S).

Now consider the lifts (relative to μ^F) of the odd coordinate derivatives $\frac{\partial}{\partial\theta}$ on the fermionic supermanifold F = $B^{0,4}$. These are precisely [AA1]

$$D_\alpha = \frac{\partial}{\partial\theta^\alpha} - (i\sigma^\mu_{\alpha\dot{\beta}})\bar{\theta}^{\dot{\beta}}\frac{\partial}{\partial x_\mu}$$

$$\bar{D}_{\dot{\alpha}} = \frac{\partial}{\partial\bar{\theta}^{\dot{\alpha}}} + \theta^\beta(i\sigma^\mu_{\beta\dot{\alpha}})\frac{\partial}{\partial x_\mu}$$

i.e. the *supercovariant derivatives.* Depending on the form of the superPoincare algebra used (e.g. Majorana spinors for Weyl spinors) alternative forms of these are possible : consult [AA1].

Since left and right multiplications on S necessarily commute, we do not have to do any work to confirm the supercovariance relation $\{Q_\alpha, D_\beta\} = 0$. Here the generators Q_α and $\bar{Q}_{\dot{\alpha}}$ of (left) fermionic translations on S have a representation amongst (right-invariant) vectorfields given explicitly by changing the signs in the second members of D_α and $\bar{D}_{\dot{\alpha}}$ above. More explicitly $\{Q_\alpha, D_\beta\}$ is an even vectorfield horizontal w.r.t. $h: S \to F$ and hence must vanish !

Notice that the supercovariant derivatives transform together as spinors under the (tangent space) action of the Lorentz group : this is a reflection of the equivariance of μ under L and the fact that the direct sum splittings in $t+(q+\bar{q})$ are ℓ-invariant.

A supersymmetry transformation (parametrised by some constant anticommuting number ε) is just left translation in S by the group element $\exp(\varepsilon Q)$: explicitly given by

$x_\mu \to x_\mu + \varepsilon(i\gamma^\mu)\theta + \bar{\theta}(i\gamma^\mu)\varepsilon$

$\theta^\alpha \to \theta^\alpha + \varepsilon^\alpha$; $\bar{\theta}^{\dot{\alpha}} \to \bar{\theta}^{\dot{\alpha}} + \bar{\varepsilon}^{\dot{\alpha}}$.

One often sees the statement that "ε is an odd spinorial parameter of the supersymmetry transformation" (cf. θ^α transformation in particular). Presumably this reflects the fact that $exp(\varepsilon Q)$ plays a dual role as group element lying in and acting on the group manifold. Since the Lorentz group does not act directly on B_1 (unless we change the Lie bracket on gB), there is no other way of justifying the phrase.

Berezin Integration.

At first sight the Berezin rules for integration on (the purely fermionic) $B^{o,q}$ appear to be rather unnatural :

$$\int 1\, d\theta^1 .....d\theta^q = 0$$

$$\int \theta^1 ....\theta^q d\theta^1 ....d\theta^q = 1$$

supplemented by "transformation via the Berezinian (= superdeterminent) under all coordinate transformations."

However, as we will show, one may combine these rules with a suitable integration on bosonic superspace and even extend to (a restricted class of) supermanifolds.

The unusual nature of the Berezin prescription is seen with q=1. If $\theta'=a\theta+b$ (a,b real) is a coordinate redefinition in $B^{0,1}$, preservation of the rule $1 = \int\theta' d\theta'$ requires $d\theta' = a^{-1}d\theta$ (the Berezinian is the inverse jacobian in fermionic space) which implies that the symbol "$d\theta$" is a contravariant object (i.e. vectorfield) rather than a covariant one. Geometrically this means that we should be considering coordinate derivatives $\frac{\partial}{\partial\theta}$ rather than coordinate one forms $d\theta$ as our basic building block.

Turning to $B^{p,q}$ with coordinates as usual $(x^1, \cdots x^p; \theta^1, ...\theta^q)$ we integrate (postulated) objects d(x;θ) that transform via the Berezinian

$$x \to \bar{x}(x;\theta) \; ; \; \theta \to \bar{\theta}(x;\theta)$$

$$d(x;\theta) \to d(x;\theta) Ber \left[\frac{\bar{x};\bar{\theta}}{x;\theta} \right].$$

The separate transformation rules for the bosonic and fermionic cases suggests that we should represent d(x;θ) by

$$d(x;\theta) = dx^1 \wedge \cdots dx^p \otimes \frac{\partial}{\partial\theta^q} \vee .... \vee \frac{\partial}{\partial\theta^1}$$

(the symbols $\wedge$ and $\vee$ will be discussed in a moment).

Unfortunately this does not transform in the way we want because under a general coordinate redefinition d(x;θ) picks up terms containing $d\theta$ and/or $\frac{\partial}{\partial x}$ - the "top" part transforming with the Berezinian. These additional terms always appear with a factor $\frac{\partial \bar{x}}{\partial \theta}$. Then the representation above does have the correct transformation property [Ro3] provided that all coordinate redefinitions are restricted by the *Z-bundle condition* $\frac{\partial \bar{x}}{\partial \theta} = 0$. Clearly one can extend the argument quite easily to supermanifolds (called Z-bundle supermanifolds hereafter) for which transition functions are all Z-bundle maps.

We return to the general case : depending on ones point of view one may view local superdiffeomorphisms in $B^{p,q}$ as coordinate redefinitions in $B^{p,q}$ or transition functions for some more general supermanifold of dimension (p,q). One might wish to consider (active point of view) the effect on integration of symmetries of flat superspace or alternatively (passive point of view) the possibility of obtaining an integration theory on some general class of supermanifolds. From the inhomogeneous nature of the transformation rule for the d(x;θ) it is natural to consider the set of all (p,q) hyperforms [PS]

$$\Omega_S^p \otimes_{H_S^\infty} \mathcal{D}^q = \mathcal{D}^q(H_S^\infty, \Omega_S^p)$$

where $\mathcal{D}$ denotes the sheaf of differential operators on S and Ω_S the sheaf of graded diferential forms. (p,q) hyperforms are just part of a larger algebra of all hyperforms [PS] $\Omega_S \otimes \mathcal{D}$. This is generated as a *right* H_S^∞-algebra by

$$dx^i \wedge d\theta^\alpha \otimes \frac{\partial}{\partial x^j} \vee \frac{\partial}{\partial \theta^\beta}$$

Here we abuse notation by considering i, α, j, β as multiindices. The (p,q)-hyperforms are generated by those generators satisfying $|i|+|\alpha|=p, |j|+|\beta|<q+1$ with $\wedge$ denoting the graded antisymmetric (wedge) product and $\vee$ the graded symmetric product. The natural double Z-grading is geometric (i.e preserved under all coordinate redefinitions).

Let $Ber(B^{p,q}) \subset \Omega^p \otimes \mathcal{D}^q(B^{p,q})$ denote the set of "top" hyperforms in the following sense. We have a new (coordinate dependent) Z-grading called the hypergrading defined on generators as follows

$$|dx^i| = 1, |d\theta^\alpha| = -1, |\frac{\partial}{\partial x^j}| = -1 \text{ and } |\frac{\partial}{\partial \theta^\beta}| = 1.$$

$Ber(B^{p,q})$ is $H^\infty(B^{p,q})$-module generated by $d(x;\theta)$. Elements of Ber consist of "top" (p+q-hypergraded) hyperforms and will be called coordinate Berezin forms. Under coordinate redefinitions coordinate Berezin forms transform simply via the Berezinian (i.e. after projection back to Ber).

On the whole algebra of hyperforms the following "Fermi integral" is well defined [Roth]

$$\int_F : \Omega_S \otimes \mathcal{D} \to \Omega_X.$$

One combines $\tilde{}: H^\infty(S) \to C^\infty(X)$ and $\Omega_S \to \Omega_X$ (kill $d\theta$ and $\frac{\partial}{\partial x}$ terms and reinterpret dx terms) with the map $\mathcal{D} \to H^\infty(S) \to C^\infty(X)$ given by $D \to [\widetilde{D(1)}]$.

e.g. The Fermi integral of a coordinate Berezin form is

$$\int_F d(x;\theta)a(x;\theta) = dx^1\wedge....\wedge dx^p \frac{\partial^q}{\partial\theta^q....\partial\theta^1} a(x;\theta)$$

$$= a_{TOP}(x)dx^1\wedge\cdots\wedge dx^p.$$

So the Fermi integral reproduces the Berezin rules on coordinate Berezin forms.

We will be considering the extension of $\int_F$ by an ordinary integration over X. In particular we seek a global formulation of Berezin integration (to answer the question : for what class of supermanifolds S can we extend the coordinate Berezin rules as above?). It is simplest to assume that the body manifold X of S is at least oriented (then it has a (Riemannian) volume form).

If U is a coordinate neighbourhood in S there is a projection $p_U:\Omega_S^p\otimes\mathcal{D}^q(U)\to Ber(U)$ onto the top hyperforms (coordinate Berezin forms) which is the local version of a global sheaf projection p which fits into an exact sequence

(*) $\quad 0\to K\to\Omega_S^p\otimes\mathcal{D}^q\underset{p}{\to}Ber\to 0$

where Ber is the Berezinian sheaf on S and we have introduced K = ker p. ["Ber" is abstractly defined by a Cech 2-cocycle "$Ber(g_{ij})$" where g_{ij} denotes transition functions on S - this making sense precisely because the Berezinian is multiplicative.]

Unfortunately * does not split naturally and there is no nice characterisation of the H_S^∞-Module K. What is known concerns

$$(\Omega^p\otimes\mathcal{D}^q)^{\rightarrow} = \{\omega\in\Omega_S^p\otimes\mathcal{D}^q : \text{given } h\in\Gamma_c(S,H_S^\infty), \omega[h]^{\sim} \text{ is exact}\}.$$

This is an H_S^∞-Module (since $\omega.h'[h]=\omega[h'h]$ and $h'h$ has compact support whenever h has compact support).

Lemma [Roth]

Suppose S admits a partition of unity with compact supports. Then

a) $K\subset(\omega^p\otimes\mathcal{D}^q)^{\rightarrow}\subset\{\omega:\int_F\omega \text{ is exact}\}$

b) $K\otimes C_X^\infty = (\Omega^p\otimes\mathcal{D}^q)^{\rightarrow}\otimes C_X^\infty$

Corollary [Roth][Ruip2]

Under the assumption that S admits a partition of unity with compact supports there is an integral $\int:\Gamma_c(S,Ber)\to R$ defined on all compactly supported Berezin forms.

Proof

If $\omega\in\Gamma_c(S,K)$, $\int_X\int_F\omega = 0$ (the integral of an exact form with compact support). The exactness of the (last arrow in the) following exact sequence presumably follows from the previous assumption about the partitions of unity :

$0 \to \Gamma_c(S,K) \to \Gamma_c(S,(\Omega^p \otimes \mathcal{D}^q)) \to \Gamma_c(S,Ber) \to 0.$

For non-compactly supported Berezin forms $\int_X \int_F$ is ambiguous via a "surface term" (i.e. integral of an exact form not necessarily with compact support) $\int_X \int_F(\omega_1-\omega_2)$ (where $P(\omega_1)=P(\omega_2)$ is the given Berezin form). See the example below.

This suggests that it might be possible to weaken the condition that Berezin forms have compact support by combining $\int$ with some sort of contour integral [Ro3,Ro4] in which the surface term vanishes by assuming that the integrand dies away fast enough. Assume that S is a Z-bundle so the body map $\varepsilon : S \to X$ factors through Z(X) and a choice of Z-bundle coordinates splits the exact sequence (*). Otherwise said the top hyperforms Ber form a geometric submodule (as opposed to quotient module) of the set of (p,q)-hyperforms. In this case one may combine the Fermi integral with contour integration **in** Z(X). Indeed $\Omega_{Z(X)}=\Omega_X$ follows from (isomorphism) $Z:C^\infty(X) \to H^\infty(Z(X))$ and $\Omega_X=\Omega C^\infty_X=\Omega H^\infty_{Z(X)}=\Omega_{Z(X)}$. This amounts to saying that $\int_F$ is given by simply killing the $d\theta$ and $\frac{\partial}{\partial x}$ terms without reinterpreting the dx terms as one forms on the body manifold.

For de Witt supermanifolds Rogers showed that "Berezin integration" is consistent on global integral forms (= (p,q)-hyperforms which are coordinate Berezin forms in each chart of *some* coordinate system) only for Z-bundles. Rogers also showed that a condition sufficient to guarantee the existence of at least one global integral form is that S is a de Witt supermanifold with a body admitting a volume form. (To be precise she showed that this is true provided that transition functions act linearly on all body map fibres which is the same thing).

We conclude with an illustrative example of how the Berezin rules can go wrong with a descripion of the correction term due to Rothstein.

Example

Consider $\varepsilon_0^{-1}((0,1)) \subset B^{1,2}$ with coordinates $(x;\theta^1,\theta^2)$. Let $\bar{x}=x+\theta^1\theta^2; \bar{\theta}^1=\theta^1+\theta^2$ and $\bar{\theta}^2=\theta^2$. This transformation (coordinate redefinition) has superjacobian $\begin{bmatrix} 1 & \theta^1 & -\theta^2 \\ 0 & 1 & 1 \\ 0 & 0 & 1 \end{bmatrix}$ and hence Berezinian 1. We easily check

$$(Ber)\int d(x;\theta)x = 0$$

$$(Ber)\int d(\bar{x};\bar{\theta})\bar{x} = (Ber)\int Ber\begin{bmatrix} \frac{\bar{x};\bar{\theta}}{x;\theta} \end{bmatrix} d(x;\theta)(x+\theta^1\theta^2) = 1$$

Certainly this coordinate Berezin form $d(x;\theta)x$ does not have compact support. Rothstein shows that, by looking in $\Omega^1 \otimes \mathcal{D}^2$ we may correct this computation as follows :

$$\int d(\bar{x};\bar{\theta})\bar{x} = \int d(x;\theta)[1-\theta^1\theta^2\frac{\partial}{\partial x}](x+\theta^1\theta^2)$$
$$= \int_0^1 -dx\frac{\partial}{\partial x}x + \int_0^1 dx.\,1$$
$$= -[x]_0^1 + 1 = 0$$

Our account emphasises that it is a "surface term" that corrects the Berezin transformation rules.

We do not intend to suggest that such inconsistencies occur in the physics literature. On the other hand if we are considering an area in which global supermanifold considerations are required we should bear in mind that Berezin integration fails to be geometric. For example, in the theory of superstrings it seems reasonable to suppose that a global theory of super Riemann surfaces (as in [CR]) will be needed. However any (Polyakov) integral over all super Riemann surfaces which incorporates a Berezin integration will run into difficulties unles one can assume that the integrand has compact support or dies away swiftly. This assumption does not appear to be valid (remark of Rabin, Crane & coworkers).

CONCLUSION

A theory of Berezin integration on a supermanifold S consists of a geometric class B of "integral (hyper)forms" (transforming via the Berezinian) on S and a linear functional on a suitable non-trivial set of global sections of B. [Other integration theories are possible including the one based on local sections [BC3] but we have not considered them here]. It appears that these conditions force S to be a Z-bundle with at least one partition of unity with compact support. It is still doubtful whether a non-trivial integration theory of Berezin type exists for any supermanifolds other than z-extended graded manifolds. One class under consideration at present is the ρ-supermanifolds of [Ho et al]. These are z-bundles because it can be shown that every ρ-supermanifold can be embedded in a de Witt supermanifold (hence there exists a partition of unity with compact support also).

Conclusion.

The anomaly in the Berezin transformation rule may (in certain cases) be corrected by looking in the space of (p,q)-hyperforms : it is seen to arise when we consider Berezin forms of noncompact support and superdiffeomorphisms that are associated with the necessity of extending the bosonic coordinates of superspace. Since Z-bundles and de Witt supermanifolds are necessarily noncompact (L>1) this warning amounts to the assertion that supersymmetry invariance in the superspace fomulation of Berezin integration is only up to surface terms and is thus only a true invariance when the integrand dies away fast enough (or has compact support).

Supermanifolds and Physics.

It is quite clear that supermanifolds should play some role in any physical theory in which superspace methods have been employed. Indeed part of the present lecture was designed to make rigorous some superspace methods. Thus supermanifolds have traditionally been employed for classical Supersymmetric field theories which have some "geometric" character. However, supermanifold theory is only likely to contribute something new when *global* supergeometry is required (i.e. there is more than one super coordinate patch). This explains why classical theories of supergravity feature so prominently in the (very subjective) list of applications below.

I) SUPERGRAVITY .

One specific problem is to give a supergeometric interpretation to the (Wess-Zumino) constraints on torsion & curvature. This theory has been attacked using bundles [DS , Y] and in conjunction with global super variational calculus [BC , Ruip]. [BC] follow the Mansouri/MacDowell approach to SG and intepret supersymmetries of the theory by comparing them with the changes in fields when two different body immersions are used. It is evident that the super vielbein approach to SG can be interpreted using suitable super frame bundles (combined with a reduction in the - tangent space - group).

II) SUSY σ-MODELS .

The problem here is to give a supergeometric interpretation of the SUSY σ-model action generalising the usual interpretation of the bosonic σ-model in terms of minimal surfaces. No really satisfactory solution to this problem has been proposed since it was shown by Rawnsley [Rawn] that the SUSY σ-model is not the action for "minimal surfaces" in graded Riemannian geometry : this is perhaps explained by the fact that Riemannian theories are torsion free whereas supersymmetric theories have torsion. Both because of its relation with string theories and because of the apparent simplicity of the superfield form of the action, the SUSY σ-model should be worthy of renewed efforts.

III) SUPERSTRING THEORIES .

String theories provide an exciting and complex set of theories and differential geometrical methods form only a small part of the technical machinery needed to understand them (along with the representation theory of Virasoro algebras, 2d conformal quantum field theory methods etc.) Since the Polyakov path integral approach to string theory explicitly "sums" over all SuperRiemann surfaces, global supermanifold techniques are required [RC]. There is a growing inductry in super Teichmuller space theory as well as the theory of super Riemann surfaces themselves [RC , BaBry]. Rabin has written about a super cohomological interpretation of the Wess-Zumino term for the covariant version of superstring theory [Ra2].

IV) QUANTISATION .

Ultimately it would be interesting to see if supermanifold techniques can be used to find a "supergeometric quantisation procedure". Apart from anything else, there are technical problems with the $L=\infty$ limit [Ro5 , Bry2] which need to be overcome. Kostant dealt with prequantisation in his original work on graded manifolds [K] which has recently received attention [N]. de Witt talked about super Hilbert space theory in his book [dW] and Rogers [Ro6] has recently attempted to find a fermionic analogue of path integration (see lectures in this volume). An interesting version of geometric quantisation [AA2] looks most promising as a precursor of supergeometric quantisation.

Research supported by the Science & Engineering Research Council, London.

REFERENCES

[AA1] V.Aldaya & J. de Azcarraga A note on the meaning of covariant derivatives in supersymmetry [JMP 26 p1818-21 1985]

[AA2] ------ Quantisation as a consequence of the symmetry group : an approach to geometric quantisation [JMP 23(7) p1297-1305 1982]

[A-GF] L.Alvarez-Gaume & D.Freedman Geometrical structure and Ultraviolet finiteness in the SUSY σ-model [CMP 80 p443-51 1981]

[Ba1] M.Batchelor Two approaches to supermanifolds [Trans A.M.S. 258 p257-70 1979]

[Ba2] ----- The structure of supermanifolds [Trans A.M.S. 253 p329-38 1979]

[Ba3] ----- "Graded manifolds and supermanifolds", p91-134 of ref [S] (an extended version of these lectures is to appear shortly in book form)

[BaBry] M.Batchelor & P.Bryant Graded Riemann Surfaces (to appear in CMP)

[BC1] U.Bruzzo & R.Cianci Structure of supermanifolds and supersymmetry transformations [CMP 95 p393-400 1984]

[BC2] ----- On the structure of superfields in a field theory on a supermanifold [Lett.Math.Phys 11 p21-6 1986]

[BC3] ----- Variational calculus on supermanifolds and invariance properties of superspace field theories [JMP 28(4) p786-91 April 1987]

[BG] C.Boyer & S.Gitler Theory of G^∞-supermanifolds [Trans A.M.S. 285 p241-67 1984]

[Bry1] P.Bryant Sheaf representations and graded manifolds (to appear in J.Pure & Appl. Alg. 1987)

[Bry2] ----- de Witt supermanifolds and infinite-dimensional ground rings (submitted to Bull.L.M.S.)

[Bry3] ----- "GH^∞-supermanifolds, DPMMS preprint 1987

[Bry4] ----- "The structure of de Witt supermanifolds", DPMMS preprint, 1987

[Bu] K.Bugajska Nature of superspace [JMP 21(8) p2091-6 1980]

[CR] L.Crane & J.Rabin "Super Riemann surfaces : uniformisation and Teichmuller theory" , EFI preprint 86-25

[CRT] R.Catenacci,C.Reina & P.Teofilatto On bodies of supermanifolds [JMP 26 p671-4 1985]

[C] R.Cianci Supermanifolds and superLie groups [JMP 25(3) p451-5 1984]

[DS] J.Dell & L.Smolin Graded manifold theory as the geometry of supersymmetry [CMP 66 p197-221 1979]

[dW] B. de Witt "Supermanifolds", CUP 1984

[F] P.Freund "Introduction to supersymmetry", CUP 1986

[G1] S.Gates,M.Grisaru,M.Rocek & W.Siegel "Superspace or 1001 lessons in SUSY" (Frontiers of physics 1983)

[G2] S.Gates,K.Stelle & P.West Algebraic origins of superspace constraints in Supergravity [Nucl.Phys. B169 p347 1980]

[GSW] M.Green,A.Schwarz & E.Witten "Superstring theory I,II" ,CUP 1987

[Ho J.Hoyos,M.Quiros,F.de Urries & J.Mittelbrunn Generalised supermanifolds I,II & III [JMP 25 p833-54 1984]

[JP] A.Jadczyk & K.Pilch Superspaces and supermanifolds [CMP 78 p373-90 1981]

[K] B.Kostant Graded manifolds, graded Lie theory and prequantisation in "Differential geometric methods in mathematical physics" (Proc. Bonn 1975) S-V lect. notes 570 p177-306 1977.

[L1] D.Leites Introduction to supermanifolds [Russ.Math. Surv. 35 p1-64 1980]

[L2] ------ Clifford algebras as superalgebras and quantisation [(transl. from Russ.) Theor.&Math.Phys.58(2) p150-2 1984]

[Lo] J.Lott Supersymmetric path integrals [CMP 108 p605-29 1987]

[MK] S.Matsumoto & K.Kakazu A note on topology of supermanifolds [JMP 27(11) p2690-2 1986]

[N] H.Nakajina Characterisation of canonical Bose-Fermi systems by "anti-Hermitian" symplectic forms [JMP 24 p1731-512 1983]

[PS] R.Picken & K.Sundermeyer Integration on supermanifolds and a geeralised Cartan calculus [CMP 102 p585-604 1986]

[Ra1] J.Rabin Berezin integration on general fermionic supermanifolds [CMP 103 p431-9 1986]

[Ra2] ----- Supermanifold cohomology and the Wess-Zumino terms of the covariant superstring action [CMP 108 p375-89 1987]

[Rawn] J.Rawnsley "Graded harmonic maps", Lecture at Trieste meeting on graded differential geometry 1985

[RC] J.Rabin & L.Crane How different are the supermanifolds of de Witt and Rogers? [CMP 102 p123-37 1985]

[Ro1] A.Rogers A global theory of supermanifolds [JMP 21(6) p1352-65 1979]

[Ro2] ----- SuperLie groups : global topology and local structure [JMP 22 939-45 1981]

[Ro3] ----- On the existence of global integral forms on supermanifolds [JMP 26(11) p2749-53 1985]

[Ro4] ----- Consistent superspace integration [JMP 26 p385 1985]

[Ro5] ----- Graded manifolds,supermanifolds and infinite dimensional Grassmann algebras [CMP 105 p375-84 1986]

[Ro6] ----- "Fermionic path integration and Grassmann Brownian motion" , preprint , King's College,London 1986 and lectures here.

[Roth] M.Rothstein "Integration on non-compact supermanifolds", Washington preprint 1985

[Ruip1] H.Ruiperez & J.Munoz Masque Graded variational calculus [J.de Math. Pures et Appl. 63 and 64 p87-194 1984]

[Ruip2] ----- An intrinsic construction of the Berezinian sheaf on a graded manifold [CR Acd Sci Paris ?]

[S] H-J.Seifert,C.J.S.Clarke & A.Rosenblum (eds) "Mathematical aspects of superspace", (NATO workshop, Hamburg 1983), Reidel Pub. Co., Dordrecht 1984

[Sch1] A.S.Schwarz Supergravity,complex structure and G-structures [CMP 87 p37-63 1982]

[Sch2] ----- (also A.Voronov) On the definition of superspace [(trans. from Russ.) Theor.&Math.Phys. 60(1) p657-63 1984]

[So] M.Sohnius Supersymmetry in superspace [Phys.Reports 128 p39-204 1985]

[Sr] P.Srivasta "Supersymmetry,superfields and supergravity", (Grad. student series in Phys), Adam Hilger 1986

[SS] A.Salam & J.Strathdee Superfields and Fermi-Bose symmetry [Phys.Rev.D11 p1521-33 1975]

[VV] V.Vladimirov & I.Volovich Superanalysis [(trans. from Russ.) Theor. & Math.Phys. 59(1) p317-34 1984]

[WB] J.Wess & J.Bagger "Supersymmetry and Supergravity", (Princeton series in physics) PUP 1983

[Y] R.Yates Fibre bundles and supersymmetry [CMP 76 255-68 1980]

BEREZIN INTEGRATION AND DIMENSIONAL REDUCTION

P D Jarvis
Physikalisches Institut, Universität Bonn
Nußallee 12, D-5300 Bonn 1, F.R.G.
and
Dept. of Physics, University of Tasmania,
Box 252C, GPO, Hobart Tas 7001, Australia

A basic theorem of elementary Berezin integration, relating integrals of the invariant $\underline{x}^2 + \bar{\theta}\theta$ to those of just the c-number part $\underline{x}^2$ in 2 dimensions lower, is discussed. Examples of such "dimensional reduction" include the explicit calculation of the surface area of a "supersphere", the critical behaviour of systemsin random external fields, and the covariant quantizaton of relativistic gauge field theories.

The use of Grassmann numbers and the concomitant technique of Berezin integration [1] has become established in the storehouse of physicists methods for describing a large variety of systems, from condensed matter physics to relativistic field theories [2]. Nevertheless there remain questions of principle and of detail regarding the mathematical constructions involved [3]. The purpose of this note is to provide some clear working examples of Berezin integrals, as usually defined, and to indicate some applications of what can be described as "dimensional reduction".

The basic mechanism involved can be stated quite simply. Consider a D/2-dimensional Berezin integral of a function of the invariant combination $\underline{x}^2 + \bar{\theta}\theta$.

$$I_{D/2} = \int d^D x d\theta\, d\bar{\theta}\;\; f(\underline{x}^2 + \bar{\theta}\theta\,). \tag{1}$$

Here $\underline{x}^2 = (x^1)^2 + (x^2)^2 + \ldots + (x^D)^2$ and we have

$$\int d\theta \;=\; 1\,, \qquad \int d\theta\, 1 \;=\; 0$$

as usual. Introducing the (D-1)-dimensional angular integral

$$\Omega_{D-1} = 2\pi^{[D/2]} / \Gamma(D/2)\,,$$

and noting that

$$f(\underline{x}^2 + \bar{\theta}\theta) = f(\underline{x}^2) + \bar{\theta}\theta f'(\underline{x}^2)$$

we have after integrating by parts,

$$I_{D/2} = -(D-2)\Omega_{D-1} I_{D-2}/2\Omega_{D-3} = -\pi I_{D-2} , \qquad (2)$$

where $I_D = I_{D/0}$ is the corresponding integral of $f(\underline{x}^2)$,

$$I_D = \int d^D x f(\underline{x}^2) .$$

Consider for example the area of a D/2-dimensional "supersphere" $z^2 + \underline{x}^2 + \bar{\theta}\theta = R^2$. The area may be defined, in analogy with the ordinary D-sphere, as

$$S_{D/2} = \int d^D x dz d\theta d\bar{\theta} \; \delta((\rho^2 + z^2)^{1/2} - R)$$

where $\rho^2 = \underline{x}^2 + \bar{\theta}\theta$ is the "super-radius". Solving the δ-function constraint for $z = \pm \sqrt{R^2-\rho^2}$ we have

$$S_{D/2} = 2 \int d^D x dz d\theta d\bar{\theta} \; \delta(z - \sqrt{R^2-\rho^2}) \frac{R}{\sqrt{R^2-\rho^2}} .$$

$$= 2 \int_{B_{D/2}} d^D x d\theta d\bar{\theta} \left| \frac{R}{\sqrt{R^2-\rho^2}} \right|$$

where the integral is over the interior of the D/2-ball, $\rho^2 \leq R^2$. In the absence of a set of "superspher ical" coordinates to complement ρ, it is natural instead to carry out the integration in Cartesian coordinates over all superspace, with a step function to pick out the superball:

$$S_{D/2} = 2 \int d^D x d\theta d\bar{\theta} \int_0^R du \; \delta(u-\rho) \left| \frac{R}{\sqrt{R^2-\rho^2}} \right|$$

Introducing conventional spherical polar coordinates $\underline{x}^2 = r^2$ etc., and noting $\rho = r + \bar{\theta}\theta/2r$ by the binomial theorem, we have

$$S_{D/2} = 2\Omega_{D-1} \int r^{D-1} dr d\theta d\bar{\theta} \int_0^R du \; \delta(u-(r + \bar{\theta}\theta/2r)) \frac{R}{\sqrt{R^2-u^2}} .$$

But the only invertible root of $u = r + \bar{\theta}\theta/2r$ is $r = u - \bar{\theta}\theta/2u$. Defining further $|x| \equiv (x^2)^{1/2}$ we have

$$S_{D/2} = 2\,\Omega_{D-1} \int d\theta\, d\bar{\theta} \int_0^R du (u(1- \bar{\theta}\theta/2u^2))^{D-1}(1- \bar{\theta}\theta/2u^2)^{-1} \frac{R}{\sqrt{R^2-u^2}}$$

$$= 2\,\Omega_{D-1} \int d\theta\, d\bar{\theta} \int_0^R du (u^{D-1} - \frac{1}{2}(D-2)\,\bar{\theta}\theta\, u^{D-3}) \frac{R}{\sqrt{R^2-u^2}}$$

In the first ($\bar{\theta}\theta$ -independent) term of the integrand we recognise the ordinary area $S_D = \Omega_D R^D$. The $\theta, \bar{\theta}$ -integration however picks out

the second term which gives

$$S_{D/2} = -\frac{1}{2}(D-2)\frac{\Omega_{D-1}}{\Omega_{D-3}} S_{D-2} \equiv -\pi\, S_{D-2}$$

in agreement with the general result. It is worthwile pointing out that a direct calculation of $S_{D/2}$ as an integral of $(\mathrm{sdet}\, G)^{1/2}$, where G is the induced metric, differs from the above by a factor of $(\mathrm{sdet}\,\eta)^{1/2}$ where η is the D/2-dimensional Cartesian metric,

$$(\mathrm{sdet}\,\eta)^{1/2} = (\det 1_D)^{1/2} \left(\det\begin{pmatrix} 0 & 1/2 \\ -1/2 & 0 \end{pmatrix}\right)^{-1/2} = 2$$

A major application of the dimensional reduction property is in the supersymmetric treatment of critical systems interacting with random external fields [4]. The generating functional for correlations in the presence of the random source may be written

$$Z[j] = \mathcal{N}\int d[\eta] \exp\{-\int d^Dx (1/2(\eta(x))^2 + j(x)\,\varphi_\eta(x))\},$$

where φ_η is the solution of the classical field equation $E(\varphi) = \eta$

$$E(\varphi) \equiv (\Delta - m^2)\varphi \quad -V'(\varphi) \quad ,$$

for the field $\varphi(x)$ describing the system coupled to the Gaussian random source $\eta(x)$. Obviously

$$Z[j] = \mathcal{N}\int d[\eta]\, d[\varphi]\; \delta(\varphi - \varphi_\eta) \exp - \int (1/2\,\eta^2 + j\varphi) \quad ,$$

and using $\delta(\varphi - \varphi_\eta) = \delta(E(\varphi) - \eta) \det \delta E/\delta\varphi$ provided φ_η is unique for given η, we can perform the η-integral and introduce auxiliary (a-number valued) fields $\bar{\psi}, \psi$ to give a formal representation of the functional determinant,

$$\int d[\bar{\psi}]\, d[\psi]\, \exp \bar{\psi} \frac{\delta E}{\delta \varphi} \psi \;=\; \det \frac{\delta E}{\delta \varphi}\,.$$

Finally

$$Z[j] \;=\; \int d[\varphi]\, d[A] d[\bar{\psi}] d[\psi] \,\exp - (S + j\varphi)$$

where A(x) is an auxiliary field introduced to give a representation of $1/2E^2$:

$$\int d[A]\, \exp\,(-\,1/2A^2 + AE) \propto \exp 1/2\, E^2\,,$$

and S is the effective action

$$S = \int d^D x \{-\,1/2A^2 + A(-\,\Delta\varphi + m^2\varphi \;+ V(\varphi)) \,+$$

$$+\; \bar{\psi}(-\,\Delta \;+ m^2 + V'(\varphi))\psi\}\,.$$

Remarkably, S can be written in an entirely supersymmetric manner,

$$S = \int d^D x\, d\theta\, d\bar{\theta}\, \{1/2\, \Phi(-\,\hat{\Delta} \;+ m^2)\Phi \;+ V(\Phi)\,\}$$

in terms of superfields

$$\Phi\,(x, \theta\,, \bar{\theta}\,) = \varphi(x) + \;\bar{\theta}\psi\,(x) \;-\; \theta\bar{\psi}(x) + \bar{\theta}\theta\, A(x)\,,$$

where $\hat{\Delta}$ is the super-Laplacian,

$$\hat{\Delta} \;=\; \Delta \;+\; \frac{\partial}{\partial\bar{\theta}}\frac{\partial}{\partial\theta}\,.$$

The D/2-dimensional superfield theory can be handled by standard superspace techniques. Introducing coordinates $X \equiv (\underline{x}, \theta\,, \bar{\theta}\,)$, $P \equiv (\underline{p}, \rho\,, \bar{\rho})$, one defines the superpropagator as

$$\Delta(X - Y) = \int dP \frac{e^{-i(X-Y)\cdot P}}{(P^2 + m^2)}$$

where "dP" stands for $d^D p\, d\rho\, d\bar{\rho}$, and where

$$X\cdot P = \underline{x}\cdot\underline{p} + 1/2(\bar{\theta}\rho \,-\, \theta\bar{\rho}\,)\,.$$

Green functions can be written out in perturbation theory in the usual way. At one loop order one has integrals of the general form

$$< \Phi(X)\,\bar{\Phi}(Y) \ldots > \propto \int dPdQ \ldots e^{-iX\cdot P} e^{-iY\cdot Q} \int dK \frac{1}{(K^2+\mu^2)^L}$$

where K is the Loop momentum and a Feynman parametrization has implicitly been introduced to bring the dependence on external momenta into the factor $\mu^2(P^2, Q^2, P\,Q, \ldots)$. The loop integration has the form of (1), and using (2),

$$< \Phi(X)\,\Phi(Y) \ldots > \propto \int dPdQ\; e^{-iX\cdot P}\, e^{-iY\cdot Q}\; I_{D-2}(P^2, Q^2, P\cdot Q \ldots),$$

where I_{D-2} is the integral for the corresponding scalar (nonsupersymmetrical) theory in (D-2) dimensions (computed using for example, dimensional regularization).

Consider now the Green functions at $\theta = \bar{\theta} = 0$ (i.e. correlations of the original field) and moreover let the spatial coordinates be restricted to a D-2- dimensional subspace, $x^{D-1} = x^D = 0$. Denoting coordinates and momenta by $X = (\tilde{x}, X^{\perp})$ etc., the dependence on the 2/2-dimensional supermomenta $P^{\perp}$, $Q^{\perp}$ ... drops out of the exponentials and results in the ordinary Fourier transform of a 2/2-dimensional superintegral:

$$< \varphi(\tilde{x}, 0)\, \varphi(\tilde{y}, 0) \ldots > \propto \int d\tilde{p}d\tilde{q}\; e^{-i\tilde{x}\cdot\tilde{p}}\, e^{-i\tilde{y}\cdot\tilde{q}}\tilde{I}_{D-2}\,,$$

where

$$\tilde{I}_{D-2} = \int dP^{\perp}dQ^{\perp}\; I_{D-2}(\tilde{p}^2 + P_{\perp}^2,\, \tilde{q}^2 + Q_{\perp}^2,\, \tilde{p}\cdot\tilde{q} + P_{\perp}\cdot Q_{\perp}\,, \ldots)\,.$$

$\tilde{I}_{D-2}$ can be reduced to the form of (1) by making an expansion in (even) powers of $P_{\perp}\cdot Q_{\perp}$ and using symmetric integration (using the superrotation invariance on the 2/2-dimensional subspace). The result (2) takes now the special form (for D = 2)

$$I_{D-2} = (-\pi)^2\, I_{D-2}(\tilde{p}^2 + 0,\, \tilde{q}^2 + 0,\, \tilde{p}\cdot\tilde{q} + 0, \ldots)\,.$$

So the (restricted) Green functions of the supersymmetric theory (and hence the correlations in the random external field) become proportional to this orderto the Green functions of the same theory formulated in D-2 rather than D, spatial dimensions. (For more details, see for example Ref. 5). In particular it is the latter model, and lower dimensionality, which determine the critical behaviour of the higher-dimensional system.

The above analysis of field theories of critical systems, involving δ-functional constraints and the appearance of integrations over c-number fields to represent functional Jacobian determinants, is very reminiscent of the formal manipulations of the path integral quantization of gauge theories. The c-number fields are the well-known Faddeev-Popov "ghost" fields [6] and there is the accompanying BRST [7] supersymmetry. Although the BRST transformations form an abelian group (rather than the superrotation group of the previous example), it is remarkable that an analogous "dimensional reduction" takes place, with respect to the count of independent physical degrees of freedom of the gauge field. For a (tensor) field transforming in a given representation of the D-dimensional Lorentz group, the rule is simply to work out the dimension of the representation in (D-2) dimensions.

As might be expected from the previous examples, covariant quantization and BRST supersymmetry can be formulated in terms of D/2-dimensional superspace (although the superfields are more constrained than in the analysis of critical systems). The counting rule reflects [8] a basic property of representations of the orthosymplectic group OSp(D/2) that their "superdimension" is just the ordinary dimension in O(D-2). The details of superfield constructions of BRST supersymmety [9] depend on which type of gauge field is involved. The issue of BRST symmetry has recently acquired considerable importance in the formulation of string field theories [10]. In closing we simply give some examples to illustrate the working of the counting rule.

As before, one consider field theory formulated over D/2-dimensional superspace. A vector gauge field $A_M(x, \theta, \bar{\theta})$ will, have components (A_μ, A_m) where $\mu = 1, \ldots, D$ and $m = D+1, D+2$ (corresponding to the Faddeev-Popov ghost fields). The superdimension is simply D-2 (correctly giving 2 independent helicity states for massless fields in 4 dimensions). For a graded antisymmetrical rank 2 tensor we would have $A_{MN} = (A_{[\mu\nu]}, A_{\mu n}, A_{(mn)})$ with superdimension $1/2D(D-1)-2D+3 = 1/2(D-2)(D-3)$, according with the usual count for these fields (equivalent to a scalar degree of freedom in 4 dimensions). For a graded symmetrical rank 2 tensor gauge field one would have $G_{MN} = ((G_{(\mu\nu)}, G_{\mu n}, G_{[mn]}$ with superdimension $1/2D(D+1) - 2D+1 = 1/2(D-2)(D-1)$. The interpretation of this result (the relevant count for the gravitational case) depends on the superfield construction. In one approach [11] a D/2-dimensional constrained vielbein gauge superfield has been shown to lead in D-dimensions to a

Jordan-Brans-Dicke scalar-tensor model.

ACKNOWLEDGMENTS

The author thanks the Alexander von Humboldt-Foundation for the award of a fellowship; members of the theory group at the Physikalisches Institut, Universität Bonn, for their hospitality, and the Australian Research Grants Scheme for finanical support.

REFERENCES

[1] F A Berezin, "The Method of Second Quantization"; Academic Press, New York (1966).
[2] See for example "Supersymmetry in Physics", eds. V A Kostelecky and David K Campbell, Physica 15D; North Holland, Amsterdam (1985).
[3] A Rogers, J Math Phys 26 (1985) 385; P Bryant, these proceedings.
[4] G Parisi and N Sourlas, Phys Rev Lett 43 (1979) 744.
[5] B McClain, A J Niemi, C Taylor and L C R Wijewardhana, Nucl Phys B217 (1983) 430.
[6] L D Faddeev and V N Popov, Phys Lett 25B (1969) 29.
[7] C Becchi, A Rouet and R stora, Phys Lett 52B (1974) 344; I V Tyutin, unpublished report FIAN 39 (1975).
[8] R Delbourgo and P D Jarvis, J Phys A16 (1983) L275.
[9] L Bonora and M Tonin, Phys Lett 98B (1981) 48; R Delbourgo, G Thompson and P D Jarvis, Phys Rev D28 (1982) 2122: N K Falck, A C Hirschfeld and J Kubo, Phys Lett 125B (1983) 175; J A Henderson and P D Jarvis, Class Quant Grav 3 (1986) L61, 4 (1987) 203.
[10] W Siegel and B Zwieback, Nucl Phys B282 (1987) 125; A Neveu and P West, Phys Lett 182B (1986) 343.
[11] P D Jarvis, Bonn HE-87-22; Class Quant Grav (submitted).

ASPECTS OF STRING THEORY

D. G. Caldi

Department of Physics, University of Connecticut

Storrs, Connecticut 06268, U.S.A.

Various topics concerning strings are discussed, starting in section I with an elementary review of the Standard Model and its problems. In section II a brief introduction to strings and superstrings is presented, especially with a view to the discussions in section III of compactification and in section IV of a geometric formulation of string theory.

I. Introduction: Review of the Standard Model and its Problems

Superstrings have been the cause of much excitement and a great deal of intense work over the past two years or so because they offer the real possibilty of a consistent, finite, quantum theory of all the fundamental interactions, including gravity. They also have afforded many opportunities for mathematics and physics to come together again, after a break of many decades, in problems of mutual current interest. Some of the topics in which the theory of superstrings has raised issues which happen to be current areas of study in mathematics or which superstrings have inspired mathematicians to take up, include the following: group theory, especially representations of Kac-Moody algebras and the related Virasoro algebras; Riemann sufaces and Teichmüller theory; complex manifolds, including

Calabi-Yau manifolds, orbifolds, and deformation theory. Even if the high hopes for ultimate unification via superstrings prove to be unrealizable, this mutually fruitful and beneficial interaction between mathematicians and physicists will remain of lasting value.

Before I can begin a discussion of strings and superstrings, it seems to me useful first to answer the question: What is everything? Although you may think you already know the answer to that one, what I mean is, everything - from the point of view of particle physics [1]. First of all, and this is a truly amazing and wonderful statement in itself, there are four (only four!) fundamental forces. All the myriad and seemingly disparate interactions we see and experience in the world about us, are all derived from just four fundamental interactions (so far as we can tell to date, and despite some recent controversy about a possible fifth force). The one longest known and studied, but also the weakest is, of course, gravity. We shall have more to say about it a bit later. The next, electromagnetism, was brought to its classical perfection in the work of Maxwell, and since then we have come to understand the electromagnetic interaction in the best tested of all physical theories, namely quantum electrodynamics. We should note in passing that electromagnetism is itself the first succesful example of what has been a powerful driving force and holy grail since Newton himself and really back to the ancient Greeks, that is, a unification, a realization that two apparently totally unrelated sets of phenomena are actually merely different manifestations of the same underlying interaction, in this case, electricity and magnetism. This exemplar has fired the imagination of physicists ever since and is, of course, the impetus for superstrings.

The phenomena for the next and last two forces were not even known until the very end of the last century and the beginning of this. The rather prosaically named weak and strong forces seemed to deal with rather esoteric matters: radioactive decay, on the one hand, and holding the nucleus together, on the other. Of course we now realize that understanding these forces is crucial to an understanding of nature on the most fundamental level. An overwhelming amount of progress has been achieved in the last two decades or so, so that we now have very well tested and universially accepted fundamental theories for these two forces. What is especially noteworthy for our concerns here, is that in the course of trying to formulate a theory of the weak interactions, one was forced to bring in electromagnetism, and so achieve another spectacular unification, the electroweak theory.

So much for the briefest of reminders about forces -- what about matter? What is called the Standard Model tells us that matter comes in three (so far) families of point-like objects, spin-1/2 fermions which are either quarks and leptons. They are:

$$\begin{pmatrix} u \\ d \end{pmatrix} \qquad \begin{pmatrix} c \\ s \end{pmatrix} \qquad \begin{pmatrix} t \\ b \end{pmatrix} \qquad \text{quarks in three colours}$$

$$\begin{pmatrix} \nu_{eL} \\ e \end{pmatrix} \qquad \begin{pmatrix} \nu_{\mu L} \\ \mu \end{pmatrix} \qquad \begin{pmatrix} \nu_{\tau L} \\ \tau \end{pmatrix} \qquad \text{leptons.}$$

(For the experts, the doublet structure here is only meant to be suggestive, and not the precise doublets for $SU(2)_L$.) There are a number of striking features to these families, for example, the repetitiveness, some of which I discuss below. I want to point out here the profound significance of the seemingly little subscript L on the neutrinos, which indicates that they are only left-handed - their spin is always antiparallel to their momentum. This is but one manifestation of the fact that the weak interactions violate parity, and consequently one must build into a fundamental theory this basic asymmetry of nature that is reflected in the chirality of the neutrino.

The way these basic elements of matter interact in the Standard Model is via exchange of spin-1 gauge fields. The interactions (excluding gravity) are constructed in analogy with quantum electrodynamics (QED), that most successful of all physical theories. Hence they are all gauge quantum field theories [2]. For QED the gauge group is U(1) so there is only one gauge field, the photon, to mediate the electromagnetic force. Gauge refers to what is called gauge invariance, which is the invarance of the Lagrangian under multiplication of the fermion fields by a *local* phase and a simultaneous compensating transformation of the gauge field. For the mathematicians, the gauge fields are connections on a principal bundle P with fibre G being the gauge group (e.g. U(1) for QED). The fermion matter fields are sections of an associated vector bundle with structure group G and a connection induced from P.

For the interaction among quarks, quantum chromodynamics (QCD), the gauge fields are called gluons. The gauge group is SU(3), and since gauge fields are in the adjoint representation, there are eight coloured gluons. A most important difference from QED is that since the group is now non-abelian, the gluons carry colour charge themselves, and so interact with themselves as well as with quarks (unlike photons which carry no charge). This makes QCD highly non-linear and very much harder to deal with. But an extremely important consequence of the non-abelian structure of QCD is that the coupling goes to zero as one probes distances going to zero. This quality of asymptotic freedom has meant that at least for short distance phenomena perturbation theory is valid. The other side of the coin, large distance behaviour, is shrouded in the murky fog of non-perturbative effects. But from our best non-perturbative tool, lattice gauge theory, we at least know that the theory almost certainly has a most desirable feature, confinement of colour charge. As for the old-fashioned strong nuclear force, we understand it to be a derived or residual force among nucleons due to the fundamental QCD force between quarks and gluons, much like the van der Waals forces of atomic and molecular physics.

The gauge group of the electroweak theory is $SU(2)_L \times U(1)$. There are three additional gauge bosons besides the photon - the W^+, W^-, and Z^0. Through a perhaps superfluous bit of slight-of-hand known as the Higgs mechanism, these three gauge bosons become massive in a process called spontaneous symmetry breaking. One of the triumphs of the theory was the prediction of these masses and the subsequent experimental discovery of these bosons right at the predicted mass values. All concerned, the theorists first, then the experimentalists were awarded Nobel prizes for their stupendous achievements. This story also illustrates an important lesson for string theorists: eventually, if string theory is to be physics, there must be experimentally testable (and one hopes in our lifetime) consequences.

So we have the Standard Model. It has the matter content mentioned above, and it has gauge fields in the adjoint representation of the gauge group $SU(3) \times SU(2) \times U(1)$. The Standard Model is standard because just about everyone in particle physics believes in it due to its many successful encounters with experiment. But this is certainly not to say that it is the whole story, nor that it doesn't have problems. It is to them that we now turn, because problems are the clues to new physics.

Probably the most basic defect of the Standard Model which renders it inappropriate for a really fundamental theory, is that it is not a true unification of the three forces it incorporates. The tensor product nature of SU(3) X SU(2) X U(1) is the weakest of associations, merely enlarging the space to accommodate all three groups. Hence, there are three independent gauge couplings. (The electroweak theory by itself also suffers from this tensor product defect, and it also is not a true and full unification.) Concomitant with this lack of true unification, goes a plethora of parameters in the model. There are masses, mixing angles, and couplings - all input parameters of the theory, to be determined by experiment, or, better, by a more fundamental theory.

There have been various attempts to solve this problem. The most explored has been the unfortunately named GUT's (Grand Unified Theories) [3]. These are based on the observation that the three gauge couplings in the Standard Model all are functions of the distance probed (or the momentum transfer q^2) - they run. And they run in such a way that if one does a lot of extrapolating (and assumes we know much more than we really do), they eventually meet somewhere in the neighborhood of 10^{15} GeV. So the idea is to pick a larger gauge group G such that SU(3) X SU(2) X U(1) $\subset$ G, and at and above energies of 10^{15} GeV, G is unbroken and there is only one gauge coupling. This then is a real unification. Among its desirable features are an explanation for electric charge quantization, and a reasonable scenario for obtaining the baryon asymmetry observed in the universe.

However, GUT's have many problems of their own. Chief among these is that they predict that the proton should decay. This is not necessarily a problem in itself, provided the proton lifetime τ_p is sufficiently long. The problem is that by now the experimental lower limit on τ_p has already surpassed the prediction of the most appealing GUT (based on SU(5)) by about two orders of magnitude. Other problems include predictions of as yet unobserved magnetic monopoles, prediction of the so-called desert between 10^3 and 10^{15} GeV in which no new physics would occur, and still too many parameters. Furthermore, one must pick the gauge group and the representations with only one's prejudice as a guide - there is no underlying principle for this as one would expect from a truly fundamental theory. All this leads me to conclude that this attempt at grand unification was premature. We will see that string theory incorporates some of the notions of GUT's, but at a higher energy scale and at the level of strings, rather than point field theories.

Let me mention here that there are also other approaches to trying to solve the problems of the Standard Model, such as technicolour and preons. However, no moderately realistic model has yet been found in these schemes.

Another problem with the Standard Model, to which we alluded above, is the repetitive generations. In the words of I. I. Rabi when learning of the discovery of the muon (the first sign of a second generation), "Who ordered that?" We don't know why there are these repeats of the basic family of up and down quarks along with the electron and its neutrino. Nor do we know how many generations there are. As is clear from Rabi's remark, this is one of the oldest questions in particle physics. We will see that strings may have something to say here.

In the Standard Model or a GUT, but even already in the electroweak theory there is a problem concerning what are called the Higgs scalar fields. These appear as fundamental scalars in the theory and are used to give masses to some gauge fields (in the spontaneously broken sector of the theory) and to the fermions. However, the masses of these Higgs particles are up for grabs; yet for the theory not to run into trouble with unitarity, they cannot be too heavy (> 1 TeV). But bosons are not like fermions which can have chiral symmetry to protect their masslessness (and then if chiral symmetry is broken, the resulting masses can be small).

A way out of this problem is supersymmetry [4], which has generated a lot of interest and work for many other reasons as well. Since there are a number of contributions at this conference dealing with various aspects of supersymmetry, let it suffice to say here for the non-experts that supersymmetry relates bosons and fermions in such a way that all particles have supersymmetric partners of the opposite statistics (for example, for the fermionic electron there is a bosonic partner, the selectron). Hence, Higgs scalars can now be light since they are also protected by chiral symmetry because their supersymmetric fermionic partners, the shiggses, are. Of course, supersymmetry is broken in our world since we do not see these partner particles with degenerate masses. Supersymmetry is also one of the central ingredients in string theory (superstrings).

Finally, no list of problems would be complete without mention of what many consider to be <u>the</u> outstanding problem in fundamental physics, namely, the cosmological constant problem or why it is so amazingly small. In field theory we subtract off the zero-point (∞) energy of fields, thus setting the energy of the vacuum to zero. However, gravity couples to this energy of the vacuum, since it couples to all energy, and the result is called the cosmological constant. The problem is that we (or astronomers anyway) can measure this E_0 or cosmological constant and it is found to be zero to astonishing precision: $E_0 < 10^{-47}$ GeV^4. Up until very recently, the solution to this problem has been only hinted at (e.g. from supersymmetry). There is now a new proposal by Coleman [5] which, though speculative, does offer at least the glimpse of a solution.

With the cosmological constant problem, we have finally gotten back to gravity, which we have been ignoring for some time in our understandable fascination with all the wonderful revelations embodied in the Standard Model, as well as with its intriguing shortcomings. But what about gravity? After all, we do have a beautiful and successful theory of gravity in Einstein's General Relativity. What more do you want? What more, is a *quantum* theory of gravity. GR is a classical theory, and so, most of us believe, it cannot be complete: fine for large-distance, macro phenomena, but incapable of giving the right answers for such micro physics problems as seen especially in cosmology and the very early universe. However, all previous attempts at formulating a quantum gravity have foundered on the treacherous rocks of the divergences resulting from the highly non-linear structure of gravitational interactions. Again it appears that strings, with their softer divergences compared to point fields, offer the real hope of a quantum gravity.

II. Strings and Superstrings

Up till now we have been talking about point field theories. For various reasons it is desirable to consider instead objects with one-dimensional extent - strings [6]. Unfortunately we cannot really describe string theory since the full formulation of the theory has not yet been accomplished. All we have are various more or less crude approximations to the real theory, but with them we have already learned a lot about strings, enough to realize that they have many beautiful and even miraculous features (which perhaps will seem less miraculous once we know what the full theory is). What we are really missing is the basic principle behind the theory, the analogue of the principle of general covariance for general relativity. As I will discuss one interesting approach to a more complete formulation of string theory in section IV, let me start here with the standard perturbative formulation based on a non-linear sigma model.

In analogy with the relativistic action for a point particle, Nambu and Goto [7] independently proposed that the action for a free relativistic string be the area of its worldsheet. Classical solutions are then extremal surfaces in spacetime. The form of the action most commonly used was popularized by Polyakov [8] but was originally written down by Brink, Di Vecchia, and Howe, and by Deser and Zumino [9]. It gives the same classical equations of motion as the Nambu-Goto action. With as usual in particle physics $\hbar = c = 1$, and the signature for the metric $G_{\mu\nu}(x)$ on the spacetime manifold M being (-, +, +, ... , +), we let (Σ, g_{ab}) be a two-dimensional spacetime manifold, the worldsheet, with intrinsic metric g_{ab}. Specializing to a closed string, Σ is isomorphic to a cylinder, $\Sigma \cong S^1 \times R$, and we can pick global coordinates (σ, τ). The action (a two-dimensional non-linear sigma model)is given by

$$S = - T/2 \int d\sigma\, d\tau \sqrt{-g}\; g^{ab}\, \partial_a X^{\mu}(\sigma,\tau)\, \partial_b X^{\nu}(\sigma,\tau)\, G_{\mu\nu}(x) \qquad (2.1)$$

(+ fermionic piece if superstring).

The parameter T is called the string tension and since it is used to make the action dimensionless, it has dimensions $[T] = [m^2] = [l^{-2}]$. The string tension sets the scale for string physics, and it was the all-important suggestion of Scherk and Schwarz [10] that this scale be given by the Planck mass, $M_{Pl} \approx 10^{19}$ GeV, and $\sqrt{T} \approx M_{Pl}$. This was a crucial step for the development of string theory as a fundamental theory of all interactions, including in particular gravity, rather than as a theory only of hadronic physics, which was what strings were originally invented for. The reason is that M_{Pl} sets the scale for quantum gravity: below this energy scale quantum fluctuations are not important in gravitational interactions, and also gravitational effects are negligible for particle interactions. Planck himself noticed that combining his constant with the speed of light c and Newton's constant G, gave one a fundamental unit of mass-energy or length: $(\hbar c/G)^{1/2} = M_{Pl} \approx 10^{-5}$ g $\approx 10^{19}$ GeV; $(\hbar G/c^3)^{1/2} = L_{Pl} \approx 10^{-33}$ cm. One should note that given this incredibly tiny distance scale, even from the point of view of a proton ($\approx 10^{-13}$ cm), a string looks like a point. This will be a key to the solution of the extra dimensions problem via their compactification.

To continue with the string action (Eq. 2.1), we notice that it is the standard harmonic map action with the map $X: \Sigma \to M$. If we vary S with respect to X^{μ}, we obtain the standard harmonic map equation (similar to the geodesic equation):

$$D^2X^{\mu} + \Gamma^{\mu}{}_{\nu\sigma}\, \partial_a X^{\nu} \partial^a X^{\sigma} = 0. \tag{2.2}$$

D_a is the covariant derivative on Σ, and $\Gamma^{\mu}{}_{\nu\sigma}$ are the Christoffel symbols associated with the metric $G_{\mu\nu}$.

Since the action S is geometric, it is invariant under diffeomorphisms both on spacetime M (i.e. coordinate changes of X^μ) and on the worldsheet Σ (i.e. coordinate changes of (σ,τ)). The latter is often called reparametrization invariance. If one views this formulation of string theory as a two-dimensional field theory on the worldsheet, then spacetime diffeomorphisms correspond to field redefinitions, and worldsheet diffeomorphisms correspond to local (gauge) symmetry transformations.

When we extremize the action with respect to the intrinsic metric on the worldsheet g_{ab} we find that the resulting energy-momentum stress tensor T_{ab} identically vanishes:

$$T_{ab} \equiv \partial_a X^\mu \partial_b X^\nu G_{\mu\nu} - 1/2\, g_{ab} (\partial_m X^\mu \partial^m X^\nu G_{\mu\nu}) = 0 . \quad (2.3)$$

Hence Eq. (2.3) is a constraint. It is also important to note that T_{ab} is automatically traceless, $g^{ab}T_{ab} = 0$, due to scale or Weyl invariance. Local rescalings of the metric g_{ab} leave S invariant, so S depends only on the conformal metric on Σ. Going to Euclidean analytic coordinates: $z = \tau + i\sigma$, $\overline{z} = \tau - i\sigma$, a local rescaling can be expressed as

$$g_{ab}(z) \rightarrow g_{ab} \exp(\varphi(z)) \quad . \quad (2.4)$$

A convenient choice is $g_{ab} = e^\varphi \eta_{ab}$ (with η_{ab} being the flat metric); this is called conformal gauge.

In our Euclidean analytic coordinates $T_{ab} = 0$ becomes $T_{zz} = T_{\overline{zz}} = 0$.

Conformal transformations (which are just analytic transformations since we are in two dimensions) $z \to f(z)$ are generated by $T_{zz} = T(z)$ (similarly, $\bar{z} \to f(\bar{z})$ is generated by $T_{\bar{z}\bar{z}} = \bar{T}(\bar{z})$). In terms of a Laurent expansion we have the famous Virasoro operators (the generators of the infinitesimal conformal group in two dimensions):

$$L_m = \frac{1}{2\pi i} \oint dz\, z^{n+1}\, T(z) \ , \qquad \bar{L}_m = \frac{1}{2\pi i} \oint d\bar{z}\; \bar{z}^{n+1}\, \bar{T}(\bar{z}) \,. \qquad (2.5)$$

Using Poisson Brackets one finds that classically the L_m 's obey the Virasoro algebra:

$$[\, L_m \,,\, L_n \,] \;=\; (m - n)\, L_{m+n} \quad . \qquad (2.6)$$

However, upon quantization in general there is an anomaly and so the algebra has a central extension:

$$[\, L_m \,,\, L_n \,] \;=\; (m - n)\, L_{m+n} \;+\; c/12\, (m^3 - m)\, \delta_{m,-n} \;\; . \qquad (2.7)$$

In order to have a conformally invariant theory, the anomaly must cancel when all contributions (including ghosts) to T_{ab} are included. The study of conformally invariant two-dimensional field theories has become something of an industry, and they are important not only for string theory but also for statistical mechanics [11].

Of importance for a later topic, let us consider a circle S^1 , and let θ

be an angular variable, $0 \le \theta \le 2\pi$, parametrizing S^1. Then an infinitesimal coordinate transformation of S^1 : $\theta \to \theta + a(\theta)$, is generated by $D_a = ia(\theta)\, d/d\theta$. So a complete basis for diffeomorphisms of the circle is

$$D_n = i\, e^{in\theta} \frac{d}{d\theta} , \tag{2.8}$$

and these satisfy (2.6). Hence the Virasoro algebra is the algebra which upon exponentiation generates the group Diff(S^1), the group of reparametrizations of the circle.

A common way to quantize in this formulation of string theory is to use the Euclidean path integral approach of Polyakov [8] in which the partition function Z is given by

$$Z = \int [dX]\,[dg]\, \exp(-S) . \tag{2.9}$$

The measure [dX] is standard for any path integral. What is new here is the measure [dg]. Polyakov proposed that one integrate over all Riemannian metrics g_{ab} on Σ and sum over all compact (for the vacuum to vacuum transition amplitude given by Z) orientable 2-manifolds (Riemann surfaces) Σ. Hence we have a topological expansion in the genus g of a Reimann surface. When one goes beyond genus one (a torus), things get complicated since one must include integration over the Teichmüller parameters, characterizing conformally inequivalent geometries. This subject has received intense investigation over the past two years, but there are still many open problems [12].

What we have been describing is of course perturbation theory and hence it has its limitations. The hope for a formalism which would allow for non-perturbative string effects is one of the main motivations for string field theory. We will discuss one recent, geometric approach to string field theory below. But another proposal is within the Polyakov formulation to find a universal, infinite-genus space to explore non-perturbative effects [13].

We now come to superstrings. First, we make everything supersymmetric so that we now have a super Virasoro algebra, super Riemann surfaces, etc. For every bosonic mode of the string, there is now also a fermionic mode. This results in many advantages, solving many of the problems which merely bosonic strings have. Chief among these is that there is no longer a tachyon in the theory. Also the critical dimension in which the theory is consistent on the quantum level is no longer $d_c = 26$, but instead $d_c = 10$ (or less). Hence we are closer to what we perceive to be our four-dimensional world.

The reader may be wondering where our friends the fundamental particles of the first section are in all this string stuff. The answer is that they are there in the vibrational modes of the string. These become upon relativistic quantization an infinite number of states of different masses and spins. Those with finite mass start at the string scale, M_{Pl}, and so decouple from lower energy physics. Although they are vitally important for the consistency and other crucial features of string theory, we will never directly observe these states. It is rather the zero-mass ground states of the string which survive to populate our low-energy world and become the familiar particles described in section I. (To be sure most of them acquire some mass along the way via symmetry breaking, but these masses are nothing compared to M_{Pl}.) Because the string incorporates all matter and gauge particles (including the graviton), it can truly unify everything. Indeed, superstrings are supposed to be candidates for a theory of everything (again an unfortunate acronym, TOE).

What generated all the revived interest in strings more than two years ago was the work of Green and Schwarz [14] showing that superstrings could be consistent quantum theories in that certain bad anomalies were cancelled but only when the gauge group of the theory was one of two related groups. The anomalies arise because it had been realized that one needed chiral fermions (i.e., left-handed multiplets different from right-handed ones) as we've seen in section I, and the only way to obtain this chiral asymmetry in the low-energy theory seemed to be to build it into the fundamental theory from the start. (Although there now seems to be a way to avoid this, it does not seem to give a realistic model [15] .) But when one has chiral fermions, the nasty possibility of gauge anomalies rears its ugly head. Indeed Green and Schwarz showed that only for the gauge group $G = E_8 \times E_8$ (or the somewhat related group SO(32)) did the anomalies cancel. Thus we had a vast improvement over the field theory GUT situation: the gauge group was not a matter of choice, but was almost uniquely predicted. Unfortunately, the picture is now not quite so rosy. As we will see, there are many different vacua to choose from (and so many different groups in $d < 10$) until we find the true dynamics governing strings.

I believe it is fair to say that still the most promising superstring is the heterotic string [16] . It is based on a closed string, but with very different left and right movers (thus building in at the start the chiral asymmetry). One formulation, the bosonic one, has the right-moving modes as those of a ten-dimensional superstring, while the left-moving modes are those of a 26-dimensional bosonic string, except that 16 of these coordinates are compactified on a 16-dimensional torus, so they are really internal coordinates. The torus is given by $T = \mathbf{R}^{16}/\Gamma$, and a vital ingredient for a consistent theory called modular invariance (related to conformal invariance) restricts Γ to be an integral, even, self-dual lattice. In sixteen dimensions there are only two: $\Gamma_8 \times \Gamma_8$ (Γ_8 is the root lattice of E_8) and Γ_{16} (the root lattice of spin(32)/Z_2). Thus again, but now from a different perspective, we have an almost unique prediction for the internal gauge group of the superstring. (Note that spin(32)/Z_2 is related to SO(32).) There is also a fermionic formulation of the heterotic string since we have a two-dimansional field theory and so can relate bosons and fermions. In this formulation there are 32 fermi fields which carry the internal symmetry.

III. Compactification

Since for low-energy physics we want to recover the Standard Model, we must get from superstrings in $d = 10$ to $d = 4$, and we must reduce the gauge group eventually to $SU(3) \times SU(2) \times U(1)$. There are now known to be a number of ways to accomplish this.

The first approach, and the one that I will concentrate on and in which there is a good realistic candidate model, is compactification on Calabi-Yau manifolds [17]. A related approach is compactification not on manifolds, but instead on spaces which have some singularities, called orbifolds [18]. By the process of blowing-up the singularities, many of these orbifolds become Calabi-Yau manifolds. There has been much interesting work here since often orbifolds are easier to calculate with. A third approach [19] is not actually compactification of a ten-dimensional string theory, but instead a formulation of the theory in $d < 10$, and especially in $d = 4$. Both bosonic and fermionic formulations are being looked at. They are generalizations of the techniques used in the original heterotic string. There are some overlapping models in all three approaches, and all also have models distinct to themselves. I should really call them vacua, since these compactifications are actually classical vacua for string theory. The problem is that there is a huge degeneracy to this space of all possible vacua. If string theory is to be truly fundamental, then eventually we must find the correct principle which picks out just the one we live in.

The basic idea of compactification is to curl up the extra six dimensions of the ten-dimensional superstring (usually heterotic string) into a "ball" of size of order L_{Pl}, hence we don't see them. I should point out that although this is related to the Kaluza-Klein idea, it is critically different in one way. The compactified space here has no continuous isometries and so generates no gauge symmetries. This is so because $E_8 \times E_8$ is already more than big enough to accomodate all known particles and gauge interactions (and then quite a bit more). Furthermore, such Kaluza-Klein gauge fields would lead to unwanted communication between the two E_8's.

The program of compactification [20] assumes that the ten-dimensional spacetime of a superstring is actually of the form $M^4 \times K$, where M^4 is four-dimensional, flat Minkowski space, and K is a compact six-dimensional space. Thus $M^4 \times K$ is supposed to be the vacuum solution to the classical string equations, or the classical background configuration for superstrings. The remakable feature of this program is that the topology and geometry of K determine so much of the resulting low-energy physics. The topological properties of K determine the low-energy gauge symmetry group and also symmetry-breaking scales. They also provide an answer to the question we discussed above, namely how many families there are and why. Remember for low-energy physics we need massless string modes and in particular for matter fields we need fermions and they are chiral. The number of generations can be shown to be given by the number of positive chirality zero modes (n^+) minus the number of negative chirality zero modes (n^-) for the Dirac operator on K: $N_G = n^+ - n^-$. But this is just the index of the Dirac operator: index $D_K = n^+ - n^-$. As a result of the Atiyah-Singer index theorem we can relate the index to a topological invariant involving the gauge field bundle F, namely:

$$N_G = \text{index } D_K = 1/2 \int_K ch_3 (F) \quad , \tag{3.1}$$

where $ch_3(F)$ is the third Chern character of F. This is a wonderful result - we finally have the framework for an answer to the generation problem, and it involves some rather fancy mathematics! Yet other things which the topology of K gives us are the masses and lifetimes of particles as determined through Yukawa couplings [21].

A class of solutions to the string equations from the non-linear sigma model is for K to be a Ricci flat Kähler manifold, a Calabi-Yau (CY) manifold. Such manifolds have SU(3) holonomy (in our d = 6 case), but no continuous isometries. With such K's one can solve in a simple way the condition for anomaly cancellation.

$$\mathrm{tr}\, R \wedge R = \mathrm{tr}\, F \wedge F\,, \tag{3.2}$$

where R is the Riemann tensor on K and F is the gauge field strength, both expressed as curvature two-forms. One embeds the holonomy group in the Yang-Mills gauge group, so that the spin connection ω (the "gauge" field of the SU(3) holonomy of K) becomes identified with some of the gauge fields A_i which form an SU(3) subalgebra. This is often called embedding the spin connection in the gauge group. Since the Yang-Mills group of interest is $E_8 \times E_8$, and one of the E_8's is assumed to describe an unknown shadow world, this results in a breaking of the E_8 responsible for our world to E_6, which is known to be relatively good for phenomenology. It is important to note that this E_6 GUT is at the level of strings. Contrary to earlier attempts, there is almost certainly no point field theory GUT [22] . Furthermore, in this scenario one can also simplify the calculation of N_G since it turns out that the right-hand side of (3.1) is now given by one-half the Euler characteristic of K, 1/2 $|\chi|$, and this is much easier to calculate.

To be accurate, CY manifolds fail to be solutions to the string equations beyond third order in sigma model perturbation theory [23] . However, one can remedy the situation by a field redefinition [24], and the problem may not even exist non-perturbatively. Furthermore, the topology is what is important for low-energy physics and it is unaffected by various quantum corrections [25] .

As already alluded to, even though there is essentially only one CY manifold in d = 4 (namely the K3 surface), there are estimated to be upwards of 100,000 in d = 6. However, as we discuss below, there is so far only one with three generations!

Let us now briefly review some of the relevant geometry and topology of CY manifolds. Some of the material on Kähler manifolds will also be useful for section IV. For more extensive treatments for physicists see refs. [6] and [17]. We start with the following definition and then try to see what it means.

Definition. A Calabi-Yau space is a compact, n-dimensional complex manifold with Ricci flat Kähler metric. For our purposes n = 3, so there are 6 real dimensions.

First, what is a complex manifold? If we have a real 2m-dimensional manifold M, then locally we can define complex coordinates as, for example:

$$\begin{aligned} z_j &= x_j + i\,x_{j+m}\,, \qquad j = 1, \ldots, m \\ \bar{z}_j &= x_j - i\,x_{j+m}\,. \end{aligned} \tag{3.3}$$

However, this is not sufficient. We need holomorphic transition functions between coordinate patches. Multiplication by i must be smooth over M. To have this we need a tensor field $J^i{}_j$ on M such that $J^2 = -1$. J is called an almost complex structure. If we define $N_{ij}{}^k = J_i{}^l\, J_{[j,l]}{}^k - J_j{}^l J_{[i,l]}{}^k$, and J satisfys $N_{ij}{}^k = 0$ (i.e. no torsion), then M is a complex manifold (providing also that M is orientable). One should note that there are many complex structures possible on a manifold. A complex structure gives a way to assign the real x-coordinates to the z-coordinates, as in (3.3), and there are many ways to do this. Also one can continuously deform one to another, so there is in fact a space of complex structures.

It is still a hard mathematical question whether a given manifold admits a complex structure, for not all orientable, even manifolds do. In fact it is now known that the only sphere S^n which does admit a complex structure is S^2, the Riemann sphere or $\mathbf{C}P^1$. Riemann surfaces in general are one-dimensional complex manifolds. The simplest example of a complex manifold is of course flat $\mathbf{C}^n$. Other interesting examples for us are $\mathbf{C}P^n$, hypersurfaces in $\mathbf{C}P^n$, and the complex torus $T^n = \mathbf{C}^n/\Gamma$.

Important topological information about a complex manifold is contained in the complex (Dolbeault) cohomology groups $H_{\bar\partial}^{p,q}(M;C)$ defined by:

$$H_{\bar\partial}^{p,q}(M;C) \equiv \frac{\bar\partial \text{ closed } (p,q)\text{-forms}}{\bar\partial \text{ exact } (p,q)\text{-forms}} \quad . \tag{3.4}$$

If the manifold has an hermitian metric, we can define the Laplacian

$$\Delta_{\bar\partial} = \bar\partial\,\bar\partial^{\dagger} + \bar\partial^{\dagger}\bar\partial \,. \tag{3.5}$$

Then a consequence of the complex Hodge theorem is that there exists a unique harmonic form $\alpha_{p,q}$ (harmonic means $\Delta_{\bar\partial}\,\alpha = 0$) which is a representative for each cohomology class $H_{\bar\partial}^{p,q}(M)$. We should also mention that the complex dimension of $H_{\bar\partial}^{p,q}(M) = h^{p,q}$, the Hodge number.

To define a Kähler manifold, we first note that on a complex, hermitian manifold one can always define the following fundamental 2-form (or, actually, (1,1)-form), also called the Kähler form:

$$K = -i\, g_{a\bar{b}}\, dz^a \wedge d\bar{z}^{\bar{b}} \,. \tag{3.6}$$

K is actually real since g is hermitian.

Definition. If dK = 0, i.e. K is closed, then g is a Kähler metric and M is called a Kähler manifold.

One of the nice properties of Kähler manifolds is that on them the complex structure J is covariantly constant with respect to the connection defined by the Kähler metric. Hence the Riemannian structure is globally compatible with the complex structure.

The simplest example of a Kähler manifold is again complex, flat n-space $\mathbf{C}^n$, with the metric $ds^2 = dz^a d\bar{z}^a$. Another simple example is the whole class of one-dimensional complex manifolds (or all Riemann surfaces). These are Kähler since dK is a 3-form, hence it is automatically closed on a 1-dim complex manifold.

There are a great many simplifications on Kähler manifolds. For example, all three Laplacians are essentially equal:

$$2\Delta_d = \Delta_\partial = \Delta_{\bar{\partial}} \,. \tag{3.7}$$

Hence, a form is harmonic if any of the Laplacians acting on it gives zero. A consequence of this is that a holomorphic form ($\bar{\partial}\alpha = 0$) on a Kähler manifold is harmonic. Another important simplification is that there are a number of relations among the Hodge numbers of a Kähler manifold, so that for the n=3 case we are interested in, there are only six independent Hodge numbers.

Working in a local coordinate system it is possible to express the Kähler metric locally (i.e. in each coordinate patch) by

$$g_{i\bar{j}} = \frac{\partial^2 F}{\partial z^i \partial \bar{z}^j} , \qquad (3.8)$$

where $F(z,\bar{z})$ is a real-valued function called the Kähler potential. Note that g is invariant if F is changed to $F(z,\bar{z}) + a(z) + b(\bar{z})$.

We finally come to Ricci Flat Kähler Manifolds (Calabi-Yau). The Ricci tensor of a complex manifold is given locally by

$$R_{j\bar{k}} = R^l{}_{jl\bar{k}} . \qquad (3.9)$$

The Ricci tensor of a Kähler manifold can be expressed in terms of the metric by

$$R_{j\bar{k}} = \frac{- \partial^2 [\log \det(g_{s\bar{t}})]}{\partial z^j \, \partial \bar{z}^k} . \qquad (3.10)$$

The Ricci form is given by

$$R \equiv i R_{j\bar{k}} \, dz^j \wedge d\bar{z}^k , \qquad (3.11)$$

and from (3.10) can be written as

$$R = -i \, \partial\bar{\partial} \, [\log \det (g_{s\bar{t}})] . \qquad (3.12)$$

It is easily seen from (3.10 - 3.12) that the real (1,1)-form R is closed, dR

= 0, on a Kähler manifold.

Furthermore, $(1/2\pi)R \in H^{1,1}(M)$, and this cohomology class depends only on the complex structure of M. What is especially significant here is that $H^{1,1}(M)$ is equal to the first Chern class, c_1, of of the cotangent bundle of M. If we write the curvature 2-form of M in a local coordinate system,

$$\Omega^i{}_j = R^i{}_{jk\bar{l}}\, dz^k \wedge d\bar{z}^l \ , \qquad (3.13)$$

then c_1 is given by

$$c_1 = (i/2\pi)\, \mathrm{Tr}\, \Omega \ , \qquad (3.14)$$

which is $(1/2\pi)R$. It is important to note that although we have expressed c_1 in terms of local coordinates, the characteristic classes deal with the global, topological properties of the bundle.

As we have just mentioned, the necessary condition that some (1,1)-form $(i/2\pi)\tilde{R}_{j\bar{k}}\, dz^j \wedge d\bar{z}^k$ be the Ricci form of some Kähler metric is that it must be closed and that it represents the first Chern class of M (this was proved by Chern). In 1957 Calabi [26] conjectured that this necessary condition is also sufficient and he proved the uniqueness of the resulting Kähler metric within each cohomolgy class and complex structure. The proof of Calabi's conjecture had to wait for almost twenty years until Yau [27] established the existence of such a metric.

Here we are interested in the particular case $R_{j\bar{k}} = 0$. Then it follows that $c_1 = 0$. But it is an obvious corollary of the Calabi-Yau theorem that the converse is also true, namely:

<u>Corollary</u>. Let M be a compact Kähler manifold with metric $g_{j\bar{k}}$ and Kähler form given by K (3.6). Let $c_1 = 0$. Then there exists a unique Kähler metric $\tilde{g}_{j\bar{k}}$ whose Ricci tensor is zero, and this Kähler metric $\tilde{g}_{j\bar{k}}$ has the same Kähler class as the original one, $g_{j\bar{k}}$. That is, the Kähler form $\tilde{K}$ is in the same cohomolgy class as K.

This is a big improvement in one's ability to construct such manifolds since it is easier to construct spaces with $c_1 = 0$, rather than to find metrics with $R_{j\bar{k}} = 0$. Indeed it is still not known how to write such a metric in closed form. However, for a discussion about approximate metrics see ref. [17]. Calabi-Yau manifolds are very difficult spaces to vizualize. They are Ricci "flat", but not actually flat, and they have no continuous isometries (for the cases of interest, namely those with SU(3) holonomy).

Of use in constructing CY manifolds is the fact that $c_1 = 0$ is equivalent to the existence of a nowhere-vanishing holomorphic (n,0)-form. This leads to an additional symmetry on the Hodge numbers,namely $h^{p,0} = h^{0,n-p}$. For our n=3 case, this reduces the six independent Hodge numbers we had for a Kähler manifold to only four for a CY manifold. It is convenient to take $h^{0,0}$, $h^{1,0}$, $h^{1,1}$, $h^{2,1}$. However, $h^{0,0}$ is the dimension of the space of constant functions, so $h^{0,0} = 1$. Furthermore, it turns out that for the Euler charteristic χ to be non-zero, $h^{1,0}$ must = 0.

Hence we are only left with <u>$h^{1,1}$ and $h^{2,1}$ as the only non-trivial Hodge numbers which characterize a CY manifold with $\chi \neq 0$.</u> The Euler characteristic is given by

$$\chi = 2 (h^{1,1} - h^{2,1}) . \tag{3.15}$$

It can be shown that $h^{1,1}$ is the number of real parameters that determine the radius and "shape" of M. Also, $h^{2,1}$ is the complex dimension of (or the number of complex parameters which determine) the space of complex structures which can be placed on M. So for the solution to (3.2) in which the spin connection is embedded in the gauge group, the number of generations is determined by $h^{1,1}$ and $h^{2,1}$.

The simplest not totally trivial example of a Ricci flat Kähler manifold is the complex three-torus T^3. However $\chi = 0$, and the holonomy group is simply the identity (hence, there are continuous isometries). Nevertheless, one can take a quotient of T^3 under some discrete symmetry group Γ, T^3/Γ. If Γ does not act freely, so that there are fixed points, then T^3/Γ is an orbifold. As mentioned above, one can still make much progress by studying strings on orbifolds [18]. One can also resolve or "blow-up" the singularities and so obtain a CY manifold.

Many of the examples of CY manifolds which have been constructed are based on algebraic varieties (which, for our purposes, we can just take to be submanifolds). Let us consider $\mathbf{CP}^n$, complex projective space in n dimensions. This is the space of n+1 complex variables z^i, minus the origin ($\mathbf{C}^{n+1} - \{0\}$), and with the identification $z^i \cong \lambda z^i$ for any complex number λ . So we have the space of complex lines in $\mathbf{C}^{n+1}$. Although these spaces are Kähler manifolds, it turns out that $h^{n,0} = 0$, so they have no holomorphic n-form and so $c_1 \neq 0$. Hence, by the Calabi-Yau theorem they are not Ricci flat. So $\mathbf{CP}^3$ does not qualifiy as a CY manifold in n=3. However, we can consider a three-dimensional submanifold of a higher dimensional $\mathbf{CP}^n$. For example, the space called $Y_{4,5}$ is the submanifold or algebraic variety of $\mathbf{CP}^4$ defined by the quintic polynomial $\Sigma (z_i^5) = 0$. This is a CY manifold, but $h^{1,1} = 1$ and $h^{2,1} = 101$, hence $\chi = -200$. So the number of families $N_G = (1/2)\,|\chi| = 100$, and this is far too many. However, one can reduce this by dividing out by discrete symmetry groups when possible. It is also unfortunate to note that the Hodge numbers for this manifold tell us that it has a 203-real-dimensional moduli space. This is rather a lot of free parameters to be determined by string dynamics.

One of the most encouraging developments in the program of superstring compactification is the discovery of a reasonably realistic model based on a CY manifold having three generations [28, 29]. The manifold of interest was discovered by Yau and Tian [28] and is in the important and large class of CY manifolds called complete intersection manifolds [30]. We start with $\mathbf{CP}^3 \times \mathbf{CP}^3$ and consider the product variety R_0 defined by three homogeneous polynomials of bidegree (3,0), (0,3), and (1,1). More explicitly, let $\{x_0, x_1, x_2, x_3\}$ be homogeneous coordinates on one $\mathbf{CP}^3$ and $\{y_0, y_1, y_2, y_3\}$ be homogeneous coordinates for the other $\mathbf{CP}^3$. Then consider the following cubic surfaces in $\mathbf{CP}^3$: $C_1 \equiv (\Sigma(x_i)=0)$, $C_2 \equiv (\Sigma(y_i)=0)$. We also need the hyperplane in $\mathbf{CP}^3 \times \mathbf{CP}^3$ defined by $H \equiv x_0y_0 + x_1y_1 + c(x_2y_2 + x_3y_3) = 0$. Then the manifold R_0 in $\mathbf{CP}^3 \times \mathbf{CP}^3$ is given by

$$R_0 = C_1 \cap C_2 \cap H \ . \tag{3.16}$$

One can check that $c_1(R_0) = 0$; however, $(1/2)\,|\chi(R_0)| = 9$, so there are too many families. In addition, the fundamental group of R_0 is trivial, $\pi_1(R_0) = 1$, so R_0 is simply connected and this won't do for symmetry breaking.

But again there are many discrete symmetries. In particular, we look at a freely acting Z_3 symmetry group G on R_0. Let $\alpha = \exp(2\pi i/3)$ so $\alpha^3 =$

1. Then G is generated by

$$g: (x_0, x_1, x_2, x_3)\times(y_0, y_1, y_2, y_3) \rightarrow$$
$$(x_0, \alpha^2 x_1, \alpha x_2, \alpha x_3)\times(y_0, \alpha y_1, \alpha^2 y_2, \alpha^2 y_3) . \quad (3.17)$$

So finally we have the quotient manifold $R \equiv R_0/G$. It has the Hodge numbers $h^{1,1} = 6$ and $h^{2,1} = 9$, so $(1/2)|\chi| = 3$, as advertised. Furthermore, $\pi_1(R_0) = Z_3$, which leads to the breaking of E_6 to $[SU(3)]^3$, i.e., close to the Standard Model.

In describing R we have picked out a particular complex structure, the one that is maximally symmetric. With this choice the Oxford group [29] has calculated many of the properties of the low-energy theory resulting from this model. They find a realistic spectrum, good inhibition of proton decay, and reasonable neutrino masses, all without unnatural adjustments. They have also calculated many Yukawa couplings [31] using sophisticated techniques from algebraic geometry (see also [32]). One of the most remarkable aspects of this manifold R is that at least within the rather large class of complete intersections, it is the unique three-generation CY manifold [33]. If we can only find the reason why nature should pick this one manifold (if it continues to pan out phenomenologically) out of all the thousands of possibilities, we'd really be in business (or perhaps out of business!).

Despite the successes of this approach, there are some drawbacks to the type of vaccum solutions we have been discussing. We've already mentioned the large number of moduli, parameters characterizing the manifold which can only be fixed by the as yet unknown string dynamics. Furthermore, it should be said that while E_6 is alright as a unifying gauge group, something smaller like SO(10) or SU(5) might be better, since then there would be fewer apparently unnecessary gauge and matter fields. For these reasons Witten [34] proposed an alternative solution to (3.2). The usual solution of embedding the spin connection in the gauge group results in a two-dimensional non-linear sigma model with what is termed (2,2) supersymmetry, meaning two positive-chirality (right-handed) and two negative-chirality (left-handed) world-sheet supersymmetries. Instead we can consider (2,0) compactifications [35, 36].

These are also based on CY manifolds M, but with deformations of the tangent bundle TM to yield two stable holomorphic vector bundles V_1 and V_2 over M. They must satisfy the topological constraints:

$$c_1(V_1) = c_2(V_2) = 0 \,, \tag{3.18}$$

$$ch_2(TM) = ch_2(V_1) + ch_2(V_2) \,, \tag{3.19}$$

where $ch_2 = (1/2)c_1^2 - c_2$ is the second Chern character. Among the matter fields of the sigma model are left-handed fermion fields which take values in V_1 and V_2 rather than in TM as in the usual (2,2) compactifications. The V_i's do not have to have structure group SU(3), so it is possible to break E_8 to SO(10) or SU(5). In addition, the number of generations N_G is now only given by (3.1) and is not equal to $(1/2)|\chi|$. Hence, it may be possible to pick a CY manifold with trivial Hodge numbers so that there would be very few parameters. So far, an interesting, but unrealistic, example of a (2,0) compactification in this approach has been found [35] which yields the gauge group SO(10). A whole new avenue has been opened up to explore in the search for the true string vacuum.

IV. A Geometric Formulation of String Theory

One of the more interesting recent developments in string theory which has the potential for a breakthrough in our understanding of non-perturbative string dynamics, is the work of Bowick and Rajeev [37] and followers [38]. It is based on the Kähler geometry of loop space. It also is based on the principles of geometric quantization [39]. The result is a classical, non-perturbative, geometric formulation of closed string field theory, with the Kähler potential of loop space being the closed string field. Despite the fact that the equations of motion for the closed string field are classical, there are quantized objects in the theory. In addition, there are also open strings in the theory, but it is not clear yet whether these are only kinematical objects needed for mathematical reasons to define some curvatures, or whether they can eventually play a dynamical role. As a warning, I should mention that there are many different spaces in this business, but I shall endeavor to keep them from getting entangled.

In order to present the general ideas of this approach, let us start with a d-dimensional Minkowski spacetime $\mathbf{R}^{d-1,1}$. (This is actually as far as explicit calculations have been done.) It is useful to consider the (∞-dimensional) space of loops on $\mathbf{R}^{d-1,1}$. This loop space $L\mathbf{R}^{d-1,1}$ is the space of maps of the circle into $\mathbf{R}^{d-1,1}$, $x : S^1 \to \mathbf{R}^{d-1,1}$. It will usually be sufficient to consider the subspace of loops at the origin, $\Omega\mathbf{R}^{d-1,1}$.

It is remarkable that not only is loop space a complex space, but it is also Kähler, as we now show [40]. Given a parameter σ on the interval $[-\pi,\pi]$, we can Fourier analyze $x(\sigma) \in \Omega\mathbf{R}^{d-1,1}$,

$$x(\sigma) = \sum_{n\neq 0} x_n \left(e^{in\sigma} - e^{in\pi} \right) , \tag{4.1}$$

with $\overline{x}_n = x_{-n}$. There is a natural choice for an almost complex structure:

$$(Jx)(\sigma) = -i \sum \mathrm{sgn}(n)\, x_n \left(e^{in\sigma} - e^{in\pi}\right) \quad . \tag{4.2}$$

This J is trivially integrable to a complex structure since the loop space of a flat space is itself flat. One can also define a closed Kähler (or also in this case, symplectic) form with Kähler potential K_0,

$$K_0 = \sum n\, x_n{}^{\mu} \overline{x}_n{}^{\nu}\, \eta_{\mu\nu} \quad . \tag{4.3}$$

The manifold with which we are actually most often concerned is not loop space itself, but instead the space of complex structures on L. This space $\mathcal{M}$ is related to the reparametrization invariance of strings. For a closed string S^1 we have seen in section II that *Diff* S^1, the group of diffeomorphisms of the circle, is the group of reparametrizations. Since the complex structure depends on the parametrization (as in (4.2)), it is clear that to preserve reparametrization invariance we should consider the space of all complex structures obtained from each other by the action of *Diff* S^1. However, it is also clear that pure rotations don't change J. Hence, the manifold $\mathcal{M}$ is given by the orbit,

$$\mathcal{M} = \mathit{Diff}\, S^1 / S^1 \quad . \tag{4.4}$$

The manifold $\mathcal{M}$ is infinite-dimensional, but it nevertheless has many nice properties. It is actually a homogeneous Kähler manifold. (For a more detailed treatment see the lectures of Professor Wess in these proceedings

[41]). Since $\mathcal{M}$ is a coset space, *Diff* S^1 acts transitively by left translations. So $\mathcal{M}$ is homogeneous, and hence we can reduce all calculations on $\mathcal{M}$ to algebraic ones on the tangent space at what we will call the origin. But here this tangent space is essentially just the Lie algebra $\underline{Diff\ S^1}$, namely our friend the Virasoro algebra (2.6). Thus there are vector fields on $\mathcal{M}$ which satisfy,

$$[L_m, L_n] = (m-n)\, L_{m+n} \ , \quad \overline{L_m} = L_{-m} \ , \quad m, n \in Z\backslash\{0\} \ . \tag{4.5}$$

Note that this Lie algebra $\underline{\mathcal{M}}$ is given by $\underline{Diff\ S^1} = \underline{S^1} + \underline{\mathcal{M}}$.

We can now write a tangent vector v to $\mathcal{M}$ at the origin as a linear combination of the L_m's,

$$v = \sum v_m L_m \ , \qquad \overline{v_m} = v_{-m} \ . \tag{4.6}$$

Then an almost complex structure $\mathfrak{J}$ on $\mathcal{M}$ is given by

$$(\mathfrak{J}v) = -i \sum \mathrm{sgn}(m)\, v_m L_m \ . \tag{4.7}$$

One can show that it is integrable near the origin [37], so $\mathcal{M}$ is a complex manifold.

It is furthermore a Kähler manifold since there is two-form ω on $\mathcal{M}$ which can be closed, $d\omega = 0$,

$$(d\omega)(L_m, L_n, L_p) = \mathcal{L}_{L_m}\omega(L_n, L_p) - \omega([L_m, L_p], L_p) + \text{cyc. perms.} \tag{4.8}$$

The first term is the Lie derivative with respect to L_m of ω, and by left-invariance it is zero. So to have $d\omega = 0$, it is necessary that

$$\omega([L_m, L_n], L_p) + \omega([L_n, L_p], L_m) + \omega([L_p, L_m], L_n) = 0 , \qquad (4.9)$$

or that,

$$(m - n)\omega(L_{m+n}, L_p) + (n - p)\omega(L_{n+p}, L_m) + (p - m)\omega(L_{p+m}, L_n) = 0 . \quad (4.10)$$

Since invariance under S^1 implys that $\omega(L_m, L_n)$ is zero unless $m + n = 0$, the most general solution to (4.10) is known [42] to be

$$\omega(L_m, L_n) = (am^3 + bm)\delta_{m,-n} , \qquad (4.11)$$

with a and b arbitrary parameters. Hence $\mathcal{M}$ admits a two-parameter family of Kähler metrics. For ω to be invertible, either (i) $a = 0$ and $b \neq 0$, or (ii) $a \neq 0$ and $-b/a \neq n^2$, $n \in \mathbf{Z}$. It turns out that choice (ii) is the interesting one for geometry.

The Riemann curvature can be computed [37, 41] to be

$$R(L_{-m}, L_n)L_{-p} = \left\{ -\theta(p - m)(p + m)^2 \left[\frac{a(p - m)^3 + b(p - m)}{ap^3 + bp} \right] \right.$$

$$\left. + (p + 2m)^2 \frac{ap^3 + bp}{a(p + m)^3 + b(p + m)} - 2mp \right\} \delta_{m,n}\delta^q{}_p L_{-q} , \qquad (4.12)$$

where m, n, p are positive. For the case $a \neq 0$, the associated Ricci form turns out surprisingly to be absolutely convergent even though it involves an infinite trace,

$$\mathrm{Ric}(L_{-m}, L_n) = \operatorname{tr}_{p>0} R^{-p}{}_{-p-mn} = \{ -(26/12)m^3 + (1/6)m \}\, \delta_{m,n} \;. \qquad (4.13)$$

What has all of this to do with strings? When we covariantly quantize a constrained system, as we've seen in section II that string theory is, then we must have ghosts, which for the bosonic string are Grassmann or, loosely, fermionic. The vacuum state of the ghosts is then a filled Dirac sea, and mathematically this is a one-dimensional (complex) vector space. However, the vacuum state changes as one moves around in the space of complex structures $\mathcal{M}$. Thus we have what is known as the anti-canonical line bundle $\overline{\Gamma}$,

$$\begin{array}{ccc} \mathbf{C} & \rightarrow & \overline{\Gamma} \\ & & \downarrow \\ & & \mathcal{M} \end{array}$$

Now we can explain the significance of the Ricci form Ric given above (4.13). One can show that Ric on the Kähler manifold $\mathcal{M}$ is the <u>curvature</u> of the covariant derivative ∇ on densities α in $\overline{\Gamma}$:

$$\nabla^2 = [\nabla, \nabla]\alpha = -\mathrm{Ric} \wedge \alpha \;. \qquad (4.14)$$

But then if $\nabla^2\alpha \neq 0$, there are no covariantly constant sections of $\overline{\Gamma}$ to identify as ghost states, and there is no reparametrization-invariant vacuum for the ghosts. So we need to do something about Ric ≠ 0.

This is where the bosonic open strings come into the picture. In the framework of geometric quantization, Bowick and Rejeev construct a bosonic Fock space for open strings, $\mathcal{B}$. This consists of holomorphic functions on loop space, i.e., functions f ox x_n for n<0 only. There exists an inner product on this space which depends (and this will be important later) explicitly on the Kähler potential on loop space,

$$\langle f, g \rangle = \int \prod_{n=1}^{\infty} dx_n d\bar{x}_n dx_0 \, e^{-K_0} \, \bar{f} \, g \quad . \tag{4.15}$$

Here we have written for the Kähler potential K_0, the one for loop space on flat spacetime. In general we have that the inner product depends on K, the Kähler potential for loop space on a general space. K will be the crucial quantity of interest.

The Fock space $\mathcal{B}$ also depends on the complex structure J of loop space, so we actually want to consider the bosonic Fock bundle $\mathcal{X}$ with fibres $\mathcal{B}$,

$$\begin{array}{ccc} \mathcal{B} & \rightarrow & \mathcal{X} \\ & & \downarrow \\ & & \mathcal{M} \end{array}$$

The bundle $\mathcal{X}$ has of course its own connection and covariant derivative, and the curvature F_0 has been computed for the flat space K_0 case with the result,

$$F_0(L_m, L_{-n}) = \{ (d/12)m^3 - \beta m \} \delta_{m,n} . \tag{4.16}$$

Thus the curvature of the bosonic string Fock bundle is just the anomaly or 2-cocycle of the Virasoro algebra. But again because the curvature F_0 does not vanish, we have an unacceptable situation since it means that the bosonic states are also not reparametrization-invariant. So neither the ghosts nor the bosonic states are gauge-invariant separately.

The solution is to combine them so that the curvature of the bosonic Fock space, F_0, cancels that of the ghost Fock bundle, Ric. Hence we construct the product bundle $\mathcal{Z}$ over $\mathcal{M}$ with fibre $\mathcal{B} \times \mathbf{C}$,

$$\begin{array}{ccc} \mathcal{B} \otimes \mathbf{C} & \rightarrow & \mathcal{Z} \\ & & \downarrow \\ & & \mathcal{M} \end{array}$$

The curvature G_{K_0} on $\mathcal{Z}$ for the flat space case is simply the sum of our previous results,

$$\begin{aligned} G_{K_0}(L_m, L_{-n}) &= F(L_m, L_{-n}) + \mathrm{Ric}(L_m, L_{-n}) \\ &= \left[\frac{d - 26}{12} m^3 + \left(\frac{1}{6} - \beta \right) m \right] \delta_{m,n} . \end{aligned} \tag{4.17}$$

So if $d = 26$, and β (the ordering parameter) = 1/6 , then $G_{K_0} = 0$, and the combined bosonic string states with ghost states are reparamterization-invariant. These states are defined by the covariantly constant sections of $\mathcal{Z}$. All of this is not new, except for giving a geometric interpretation to the critical spacetime dimension (26 for bosonic strings) and the conformal anomaly.

We get something new when we go to a general loop space (or just a general infinite-dimensional Kähler manifold) with Kähler potential K. Bowick and Rajeev propose that since K is a real scalar function on loop space, it is the dynamical variable for <u>closed</u> strings - K is the closed string field. Its equation of motion is the generaliztion of (4.17) to general K,

$$G_K(L_m, L_{-n}) = F_K(L_m, L_{-n}) + \mathrm{Ric}(L_m, L_{-n}) = 0 . \qquad (4.18)$$

We require that the curvature G_K on the generalized combined bundle $\mathcal{Z}$ must vanish for reasons similar to those above. Note that Ric is the same as before (4.13) since it does not depend on K, only F_K does. Eq. (4.18) is a non-linear, non-perturbative equation of motion for the interacting closed string field. It is the analogue of Einstein's equation for general relativity. But note that we don't yet have a Lagrangian from which to derive this equation of motion. This is currently under investigation.

One of the encouraging results of this program is that in the low-energy limit, Bowick and Rajeev have been able to recover the full, non-linear Einstein equations in the adiabatic approximation (i.e., gravitational field slowly varying over the size of a loop). This is a definite improvement over the results of usual string theory, where only the linearized Einstein equations are recovered in the weak field approximation.

There is, to be sure, much further investigation needed before one can say that this formulation is consistent and complete, or even that it makes sense. Already there have been generalizations to superstrings [38], and there has been some speculation that this approach may be connected to the infinite genus Riemann surface program [13]. At any rate, it is a fascinating proposal, bringing together some beautiful mathematics, especially geometry, and fundamental physics - one of the main themes of these lectures.

Acknowledgement

It is a pleasure to thank M. Bowick, S. Rajeev, J. Schwarz, B. Greene, G. Ross, and W. Abikoff for stimulating and enlightening discussions. The author is a U. S. Department of Energy Outstanding Junior Investigator, whose work is supported in part by the U. S. Department of Energy under Contract No. DE-AC02-79ER10336. He is also grateful to his colleagues at the Physics Department at Yale University, where he is a Research Affiliate, for many valuable discussions.

References

1. See, e.g., K. Gottfried and V. F. Weisskopf, *Concepts of Particle Physics*, Vols. I & II (Oxford, 1986).
2. See, e.g., L. D. Faddeev and A. A. Slavnov, *Gauge Fields* (Benjamin/Cummings, London, 1980).
3. See, e.g., P. Langacker, Phys. Rep. **72** (1981) 185; G. G. Ross, *Grand Unified Theories* (Benjamin, San Francisco, 1985).
4. See, e.g., J. Wess and J. Bagger, *Supersymmetry and Supergravity* (Princeton, 1983).
5. S. Coleman, Harvard preprint, HUTP-88/A022, April, 1988.
6. M. B. Green, J. H. Schwarz, and E. Witten, *Superstring Theory* (Cambridge, 1987); for a reprint collection, see *Superstrings*, edited by J. H. Schwarz (World Scientific, Singapore, 1985).

7. Y. Nambu, Lectures at the Copenhagen symposium (1970); T. Goto, Prog. Theor. Phys. **46** (1971) 1560.
8. A. M. Polyakov, Phys. Lett. **103B** (1981) 207 and 211.
9. L. Brink, P. Di Vecchia, and P. Howe, Phys. Lett. **65B** (1976) 471; S. Deser and B. Zumino, Phys. Lett. **65B** (1976) 369.
10. J. Scherk and J. H. Schwarz, Nucl. Phys. **B81** (1974) 118.
11. A. A. Belavin, A. M. Polyakov, and A. B. Zamolodchikov, Nucl. Phys. **B241** (1984) 333; D. Friedan, Z. Qiu, and S. H. Shenker, Phys. Rev. Lett. **52** (1984) 1575.
12. For a review see, e.g., L. Alvarez-Gaumé and P. Nelson, Cern preprint, CERN-TH.4615/86, December,1986.
13. D. Friedan and S. Shenker, Phys. Lett. **175B** (1986) 287; Nucl. Phys. **B281** (1987) 287.
14. M. B. Green and J. H. Schwarz, Phys. Lett. **149B** (1984) 117; Nucl. Phys. **B225** (1985) 93.
15. L. Dixon, V. Kaplunovsky, and C. Vafa, SLAC-Harvard preprint, SLAC-PUB 4282, HUTP-87/A034 (1987).
16. D. J. Gross, J. A. Harvey, E. Martinac, and R. Rohm, Nucl. Phys. **B256** (1985) 253; **B267** (1986) 75.
17. For a review, see D. G. Caldi in *Proceedings of the International School on Relativity, Supersymmetry and Strings, Utah State University* (Plenum, New York, 1987).
18. L. Dixon, J. Harvey, C. Vafa, and E. Witten, Nucl. Phys. **B261** (1986) 678; **B274** (1986) 285.
19. H. Kawai, D. C. Lewellen, and S.-H. H. Tye, Nucl. Phys. **B288** (1987) 1; R. Bluhm, L. Dolan, and P. Goddard, Nucl. Phys. **B289** (1987) 364; I. Antoniadis, C. Bachas, and C. Kounnas, Nucl. Phys. **B289** (1987) 87.
20. P. Candelas, G. Horowitz, A. Strominger, and E. Witten, Nucl. Phys. **B258** (1985) 46.
21. A. Strominger and E. Witten, Comm. Math. Phys. **101** (1985) 341; A. Strominger, Phys. Rev. Lett. **55** (1985) 2547.
22. M. Dine and N. Seiberg, Phys. Rev. Lett. **55** (1985) 366; V. Kaplunovsky, Phys. Rev. Lett. **55** (1985) 1036.
23. M. T. Grisaru, A. E. M. Van de Ven, D. Zanon, Phys. Lett. **173B** (1986) 423; D. J. Gross and E. Witten, Nucl. Phys. **B277** (1986) 1; M. D. Freeman and C. N. Pope, Phys. Lett. **174B** (1986) 48.
24. D. Nemeschansky and A. Sen, Phys. Lett. **178B** (1986) 365; L. Witten and E. Witten, Nucl. Phys. **B281** (1987) 109.
25. E. Witten, Nucl. Phys. **B268** (1986) 79; M. Dine, N. Seiberg, X. G. Wen, and E. Witten, Nucl. Phys. **B278** (1987)

769; **B289** (1987) 319;
J. Distler and B. Greene, Cornell-Harvard preprint, CLNS 88/834, HUTP-88/A020, March, 1988.

26. E. Calabi, in *Algebraic Geometry and Topology: A Symposium in Honor of S. Lefchetz* (Princeton,1957).
27. S.-T. Yau, Proc. Natl. Acad. Sci. **74** (1977) 1798; Comm. Pure Appl. Math. **31** (1978) 339.
28. S.-T. Yau, in *Proceedings of the Argonne Symposium on Anomalies, Geometry and Topology,* W. A. Bardeen and A. R. White, eds. (World Scientific, Singapore, 1985).
29. B. Greene, K. Kirklin, P. Miron, and G. G. Ross, Phys. Lett. **180B** (1986) 69; Nucl. Phys. **B278** (1986) 667; Nucl. Phys. **B292** (1987) 606; see also, J. Ellis, K. Enqvist, D. V. Nanopoulos, and K. Olive, Phys. Lett. **188B** (1987) 415;
S. Kalara and R. N. Mohapatra, Maryland preprint, UMN-TH-598-87.
30. P. Griffiths and J. Harris, *Principles of Algebraic Geometry* (Wiley, New York,1978).
31. B. R. Greene, K. H. Kirklin, P. J. Miron, and G. G. Ross, Phys. Lett. **192B** (1987) 111;
J. Distler, B. Greene, K. Kirklin, and P. Miron, Harvard preprint, HUTP-87/A024, March, 1987.
32. P. Candelas, Texas preprint, UTTG-05-87.
33. P. S. Aspinwall, B. R. Greene, K. H. Kirklin, P. J. Miron, Harvard preprint, HUTP-87/A030;
P. Candelas, A. M. Dale, C. A. Lütken, and R. Schimmrigk, Cern preprint, CERN-TH.4694/87, UTTG-10-87, March, 1987.
34. E. Witten, ref. 25.
35. J. Distler, Phys. Lett. **188B** (1987) 431;
J. Distler and B. Greene, Harvard preprint, HUTP-87/A065, August, 1987.
36. M. Cvetic, Pennsylvania preprints, 1987;
M. Dine and N. Seiberg, IAS preprint, IASSNS/HEP-88/3;
M. Dine and C. Lee, IAS preprint, IASSNS/HEP-87/70.
37. M. J. Bowick and S. G. Rajeev, Phys. Rev. Lett. **58** (1987) 535; Nucl. Phys. **B293** (1987) 348; MIT preprint, CTP# 1494 (July,1987) to appear in Proceedings of the Eleventh Johns Hopkins Worshop on Current Problems in Particle Theory (World Scientific, Singapore, 1987).
38. K. Pilch and N. Warner, MIT preprint, CTP# 1457 (1987), to appear in Class. Quantum Grav. ;
D. Harari, D. K. Hong, P. Ramond, and V. G. J. Rodgers, Nucl. Phys.

B294 (1987) 556.
J. Mickelsson, MIT preprint, CTP# 1448 (1987), to appear in Comm. Math. Phys.

39. See, e.g., N. J. M. Woodhouse, *Geometric Quantization* (Oxford, 1980).
40. For related mathematics see A. M. Pressley and G. B. Segal, *Loop Groups* (Oxford, 1986).
41. J. Wess, these proceedings; B. Zumino, Berkeley preprint, LBL-24319, UCB-PTH-87/48 (November, 1987), to appear in 1987 Cargèse Summer School Proceedings.
42. G. Segal, Comm. Math. Phys. **80** (1981) 301.

FIELD AND STRING QUANTIZATION IN CURVED SPACE-TIMES

Norma SANCHEZ

Observatoire de Paris, Section de Meudon
92195 Meudon Principal Cedex
FRANCE

Abstract : We discuss string theory in the context of quantum gravity. We report on our recent results :

(i) (first) string quantization in Rindler space-time and its application to the Schwarzschild geometry ;

(ii) the Hawking-Unruh effect in string theory and its description in terms of analytic mappings ;

(iii) a new approach to string quantization in curved space-time in order to take into account strong curvature effects of the geometry.

CONTEXTUAL BACKGROUND

Perhaps the main challenge in theoretical physics today is the unification of all interactions including gravity. At present, string theories appear as the best candidates to achieve such an unification. However, many technical and conceptual problems remain and a quantum theory of gravity is still non available. Continuous effort over the last quarter of a century has demonstrated the many difficulties encountered in repeated attempts to construct such a theory and have also indicated some of the particular properties which an eventual complete theory will have to possess. The amount of work in that direction can be by now presented in two different sets which have most evolved (and remain) separated : (i) **conceptual unification** (introduction of the uncertainty principle in general relativity, the interpretation problem, Q.F.T. in curved space time and by accelerated observers, Hawking radiation and its consequences, "wave function of the universe",...) (ii) **grand unification** (the unification of all interactions including gravity from the particle physics point of view. Gravity is considered as a massless spin two particle (the graviton) : supergravities, Kaluza-Klein theories and the more successful superstrings).
Whatever the final theory of the world will be if it is to be a theory of everything, we would like to know what new understanding it will give us about the singularities of classical general relativity. If string theory would provide a theory of quantum

gravity, it should give us a proper theory (not yet existent) for describing the ultimate state of quantum black holes and the initial (say, very early) state of the Universe. That is, a theory describing the physics (and the geometry) at Planck energies and lengths. Till now gravity has not completely been incorporated in string theory : strings are most frequently formulated in a flat space-time. Gravity appears through massless spin two particles (graviton). One disposes only of partial results for strings in curved backgrounds ; these mainly concern the problem of consistency (validity of quantum conformal invariance) through the vanishing of the beta functions. The non-linear quantum string dynamics in curved space-time have only been studied in the slowly-varying approximation for the geometry (background field method) where the field propagator is essentially taken as the flat space Feynman's propagator. Clearly, such approximations are useless for the study of strings in strong curvature regimes where quantum gravity effects are important. Our aim is to properly understand strings in the context of quantum gravity. As a first step in this program we propose ourselves to study Q.S.T. (Quantum String Theory) in curved spacetimes. (Of course the main goal should be to extract the particle spectrum and the space time itself from string theory but, as it is known, one is very far from doing that explicitely). There are different kind of effects to be considered here : **ground state and thermal effects** (associated to the fact that in General Relativity there are no prefered reference frames, and to the possibility of having different choices of time) and **curvature effects** which will modify the mass formula, critical dimension and scattering amplitudes of strings.

FIELDS IN CURVED SPACE

The formulation of Quantum Field Theory in non trivial (curved or flat) space times has given new fundamental features with respect to the usual understanding of Q.F.T. in trivial (Minkowski flat) space time, i.e. : i) the possibility for a given field theory to have different alternative well defined Fock spaces (different "sectors" of the theory) ; (ii) the presence of "intrinsic" statistical features (temperature, entropy) arising from the non-trivial structure (geometry, topology) of the space time and not from a superimposed statistical description of the quantum matter fields. Relevant examples are Q.F.T. on the Rindler manifold and its analytic mapping extensions [1-2], black-holes and cosmological (de Sitter) space-times.

Quantum field theory developed for curvilinear (accelerated) coordinates in flat space in a way which can be directly generalized to curved space-time is an useful step for a physical and mathematical discussion of the full theory. The genuine coordinate independence which is so familiar in the classical theory of general relativity is not a particular property of gravity but a fundamental principle prevalent in all descriptions of physical laws. On the other hand, the apparent difference which results from the treatment of a quantum field theory in a variety of coordinate systems (in either curved or flat space-time) is not a coordinate effect at all, but is a consequence of the fact that physically different quantum states are correctly described by the quantum theory as being physically distinct. "Canonical" states for different coordinate systems are physically different (each timelike vector field leads to a separate indication of what constitutes a definition of positive frequency).

Some time ago the present author (2) proposed a new approach to Q.F.T. in curved space times based on analytic (holomorphic) mappings. The mappings relating some manifold to its global analytic extension are an essential ingredient in the discussion of Q.F.T. on curved manifolds and its thermal properties. This approach allows

(i) study the vacuum and thermal properties of the quantum theory enterely in terms of the properties of holomorphic functions

(ii) classify the vacuum spectra (particle production rates and their associated temperature (and entropy) according to the singularities of the mappings (the singularities here describing the asymptotic regions of the space time).

Consider a two dimensional space-time as embedded in a higher four dimensional one with metric tensor g. If g has removable singularities (event horizons), the coordinates (without horizons) in which the manifold has its maximal analytic extension can be found from mappings of the form

$$x_k \pm t_k = f\,(r^* \pm t)$$

where (x_k, t_k) are Kruskal (maximal) type coordinates and (r^*, t) are of Schwarzschild type. For the most important metrics in General Relativity, the presence of isometry groups allows to perform the maximal analytic extension of the 4-dimensional manifold M through the extension of a relevant two-dimensional manifold containing the time axis and a suitable spatial coordinate. This two-dimensional manifold can be taken as the fibre of the manifold M considered as an appropriate fibre bundle. Examples of this situation are the Taub-NUT (T-NUT) and Kerr-Newman (K-N) metrics which can be also generalized to be solutions of the Einstein equations with non-zero cosmological term, i.e. the Taub-NUT-de Sitter and Kerr-Newman-de Sitter families. After analytic continuation $t_k \blacksquare i\tau_k$ $(t = i\tau)$, the mapping

$$u_k = f(u')$$

$(u_k = X_k \pm i\tau_k\,,\ u' = r^* \pm i\,\tau)$ is holomorphic (or anti-holomorphic). These mappings should satisfy the boundary conditions

$$u_{k\,\pm} = f\,(\pm\infty)$$

$u_{k\pm}$ are the images of the event horizons in u'. Thus, in Schwarzschild spacetime $r^* \equiv r + r_s\ \ell n\,(\frac{r}{r_s} - 1) = -\infty$, (ie $r = r_s \equiv$ Schwarzschild radius) corresponds to $u_{k\pm} = 0$ (ie $x_k = \pm\, t_k$). For the Kerr-Newman or Taub-NUT families, f is the <u>exponential</u> <u>mapping</u>, ie

$$f(u') = e^{2\pi\beta^{-1}u'}$$

where

$$\beta = \begin{cases} 8\pi\left[M + M^2 (M^2 + \ell^2 + Q^2)^{-1/2} \right] & \text{(K-N)} \\ \pi\left[\dfrac{M(M+1) + 2(L^2 - M\sqrt{M^2+L^2})}{L\sqrt{M^2+L^2}}\right] & \text{(T-NUT)} \end{cases}$$

The temperature (T) of these solutions is a topological invariant : it is given by the differential winding number of the mapping f, ie

$$T = \frac{1}{2\pi} \frac{\partial}{\partial\tau} \operatorname{Im} \log f(r^* + i\tau)$$

If f is the exponential mapping, τ is periodic ($0 \leq \tau \leq \beta$) and $T = \beta^{-1}$ is the Hawking temperature. In the Euclidean regime, these solutions exist as gravitational instantons (ie complete solutions of the Einstein equations with ++++ signature). The euclidean action $I = \beta^2/16\pi$ is finite, interpreted as the intrinsic entropy of the solutions and related to its Euler number. (The presence of event horizons is a necessary condition to have non-zero entropy and a true (global) thermal equilibrium over the whole spacetime but they are not necessary to have finite temperature and asymptotic thermal equilibrium).

STRINGS IN RINDLER SPACE

Much of the present interest in string theories comes from the hope that they may provide a finite theory of quantum gravity. It seems natural in this context to investigate the quantization of strings in a curved space-time. As a first step in this programme we study the quantization of a string in a uniformly accelerated space (D-dimensional Rindler space). Although this is a flat space-time, it possesses a space-time structure including an event horizon, similar to a black-hole manifold. In order to quantize the string properly in this manifold, a horizon regularization is needed. This regularization can be introduced through the definition of the Rindler co-ordinates as follows:

$$x_1 \pm x_0 + \varepsilon = e^{\alpha(X^1 \pm X^0)}$$

$$x^i = X^i, \quad i = 2, \ldots, D \tag{1}$$

where (x^1, x^0) and (X^1, X^0) are respectively Minkowskian (Kruskal-like) and Rindler (Schwarzschild-like) co-ordinates. α is a constant defining the proper acceleration of the Rindler observers (α is equal to the surface gravity in the black hole case) and ε is an infinitesimal parameter of the order of the Planck length. The horizon is now at a finite distance $|X^1 \pm X^0| \sim 1/\alpha \ \ln 1/\varepsilon$. This regularization reflects the fact that a classical description of the geometry is no longer valid at distance of the order of the Planck length.

We quantize the string in a light-cone gauge where the light-cone variables are $(x^1 \pm x^0)$ and $(X^1 \pm X^0)$. This choice is particularly convenient here since the acceleration points in the x^1 direction. We recall that for Q.F.T. in accelerated frames (in flat and curved space-times), different time- like vector fields lead to inequivalent positive frequency modes[1)-3)]. In the light cone gauge, positive frequency modes with respect to the time-like variable (τ) on the string world sheet are physical states when τ is identified with the appropriate time variable in the physical space-time. We associate inertial and accelerated particle states of the string to positive frequency modes with respect to x^0 and X^0 respectively. We find that the accelerated frequencies of the accelerated modes differ in a large factor

$$\lambda_0 = \frac{2\pi\alpha}{\ln\left(\frac{2\pi}{\varepsilon} + 1\right)} , \qquad (2)$$

from the inertial ones. Physically this factor reflects the indefinite increasing of the string length when it approaches the event horizon (Fig. 1). We develop the string dynamics in Rindler space and write the corresponding constraints. The longitudinal co-ordinates X^0 and X^1 obey non-linear equations of motion but, as in the inertial case, they can be eliminated in terms of the transverse co-ordinates which are the independent dynamical variables. Two possible situations appear depending on whether the centre of mass has a uniform speed or a uniform acceleration. The mass formulae are derived in each case. The Poincaré invariance of the flat Minkowski space-time has a non-linear realization in terms of the Rindler co-ordinates. We prove that the passage from the inertial to accelerated modes of the string is a canonical transformation both at the classical and quantum levels. We also check that an explicit realization of the Poincare algebra can be constructed in terms of the accelerated modes, provided (at the quantum level) one sets D = 26 and uses symmetric ordering of the operators. The explicit expression of the Bogoliubov coefficients relating the

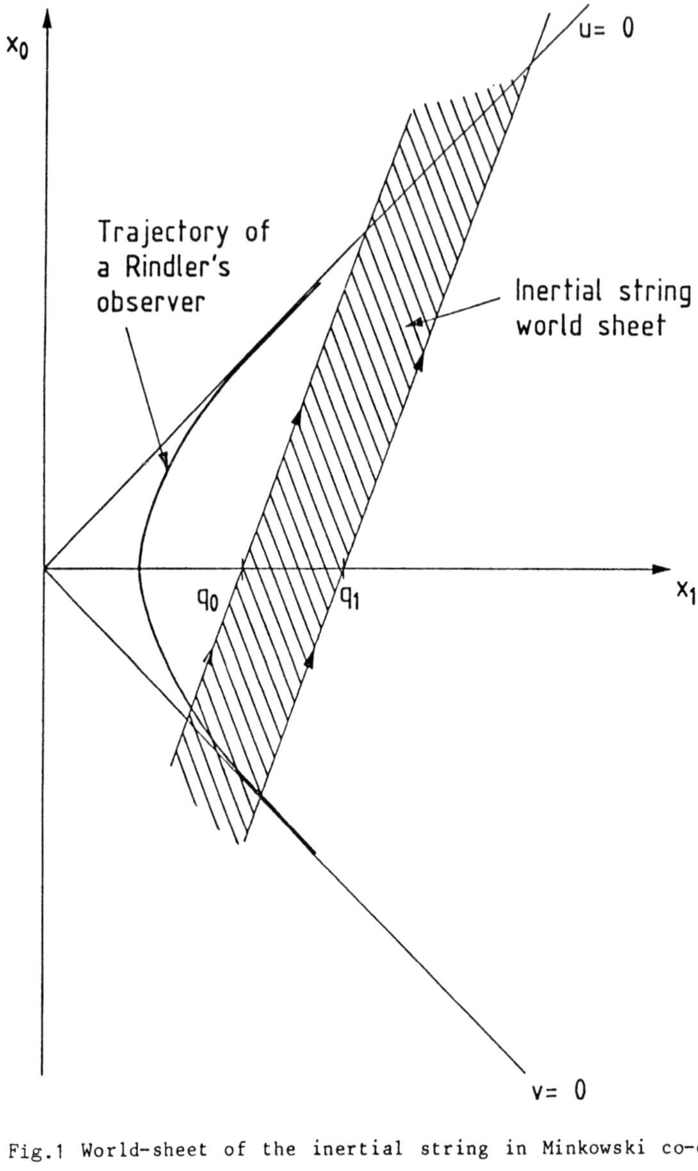

Fig.1 World-sheet of the inertial string in Minkowski co-ordinates. u = 0 and v = 0 are the horizons of Rindler space.

inertial and Rindler modes of the string is found. The expectation value of the Rindler number mode operator in the ground state (tachyon) of the string is computed. This follows a thermal distribution [Eq. (62)] with temperature

$$T_s = \frac{\alpha}{2\pi} \tag{3}$$

This is the same Hawking-Unruh value that appears in the field-theoretical context. However, if one measures the frequency in dimensionless units (1, 2, ...) instead of multiples of λ_0 [Eq. (2)], the temperature of the ground state is a (very large) pure number

$$T_0 = \frac{T_s}{\lambda_0} = \frac{1}{4\pi^2} \ln\left(\frac{2\pi}{\varepsilon} + 1\right) \tag{4}$$

We find the expectation value of the accelerated mass M' operator in the (tachyon) ground state; it turns out to be a large positive number

$$M' = 2\pi T_0 \tag{5}$$

The major features of the string quantization in Rindler space also holds for a string in a Schwarzschild space-time. An appropriate light-cone gauge can be introduced for left or right movers in the null Kruskal and Schwarzschild co-ordinates. Now, Eq. (2) for λ_0 gives the relation between the Kruskal and Schwarzschild frequencies and T_s and T_0 [Eqs. (3)-(4)] give the temperatures characterizing the ground state of the string in the black hole manifold with

$$\varepsilon \sim \left(\frac{M_{Pl}}{M}\right)^{\frac{D-2}{D-3}} \quad \text{and} \quad \alpha \equiv K \sim \left(\frac{M_{Pl}}{M}\right)^{\frac{D-2}{D-3}}$$

Here M_{pl} and M stand for the Planck mass and the black hole mass respectively. It can be noticed that the Hagedorn temperature in this context ($1/\sqrt{\alpha'} \sim M_{pl}$) is always $\gtrsim T_s \sim M_{pl}(M_{pl}/M)^{1/(D-3)}$ since a basic requirement for the present semiclassical treatment is $M \gtrsim M_{pl}$. The investigations presented here can also be extended to the case of fermionic strings and to more general (non-uniform) accelerations described by analytic (holomorphic) mappings as in the approach of Ref. 2).

HORIZON REGULATION IN RINDLER SPACE - INTRODUCTION OF THE ε PARAMETER

Let us discuss now some features of Rindler space which are important for our study of strings in this space. The Rindler transformation maps the right-hand wedge $x^1 > |x^0|$ of Minkowski space onto the whole Rindler space $-\infty < X^1, X^0 < +\infty$ (the whole Minkowski space can be covered using four different Rindler patches). As is known, a quantum field in Rindler space is in a thermal state with temperature $T = \alpha/(2\pi)$. In addition, ultra-violet divergences arise in the free energy and entropy of quantum fields from the existence of a horizon at $x^1 = |x^0|$ (i.e. $X^1 = -\infty$) in the space-time. The same problem appears in the case of a four-dimensional black hole. Let us illustrate this phenomenon by considering a free massive scalar field Ψ. In D-dimensional Rindler space, the positive frequency modes are

$$\Psi = \frac{1}{(2\pi)^{\frac{D-2}{2}}} \, e^{i(-\lambda X^0 + k^i X^i)} \, \Phi(X^1), \quad \lambda > 0, \; k^i \in \mathbb{R} \tag{6}$$

where $\phi(X^1)$ satisfies

$$\left[\frac{d^2}{dX^{1\,2}} + \lambda^2 - (m^2 + k^{i\,2})\alpha^2 e^{2\alpha X^1}\right] \Phi(X^1) = 0$$

The total number ($\mathcal{N}$) of wave modes with frequency less than λ can be computed in the semi-classical approximation for $\lambda \gtrsim \alpha$. This is enough to study the ultra-violet behaviour of the quantities interesting us. In the W.K.B. approximation $\mathcal{N}_\lambda$ is given by

$$\pi \mathcal{N}_\lambda = \int_{-H}^{a(\lambda)} dX^1 \int \prod_{i=1}^{D-2} \frac{dk^i}{2\pi} \sqrt{\lambda^2 - (m^2 + k^{i\,2})\alpha^2 e^{2\alpha X^1}} \tag{7}$$

The integration is taken over the values of k^i and X^1 for which the argument of the square root is positive. Here $a(\lambda)$ is the classical turning point

$$e^{\alpha a(\lambda)} = \frac{\lambda}{m\alpha},$$

and H is a large cut-off ($H \gtrsim 1/\alpha$) on the negative Rindler co-ordinate X^1. This shifts the horizon by replacing the light-cone $x^1 = |x_0|$ as a boundary of Rindler space-time by the hyperbola

$$(x^1)^2 - (x^0)^2 = e^{-2\alpha H} \tag{8}$$

This regularization takes into account the fact that a classical description of the geometry is no longer valid at distances of order of the Planck length. Thus,

$$e^{-2\alpha H} \sim l_P$$

Evaluating $\mathcal{N}_\lambda$ for large H it yields

$$\pi \mathcal{N}_\lambda = \frac{1}{4} \frac{(\lambda/\alpha)^{D-1}}{(4\pi)^{(D-3)/2}\,(D-2)} \, e^{\alpha H (D-2)} \left[1 + O(e^{-2\alpha H})\right] \tag{9}$$

Then the free energy and the entropy at temperature T:

$$F = -\int_0^{\infty} \mathcal{N}_\lambda \frac{d\lambda}{(e^{\lambda/T} - 1)} \quad , \qquad S = -\frac{\partial F}{\partial T} \, ,$$

are equal to

$$F = -\frac{\Gamma(D)\ \zeta(D)\ T}{(4\pi)^{(D-1)/2}\ (D-2)} \, \alpha^{1-D} \, e^{\alpha H (D-2)} \left[1 + O(e^{-2\alpha H})\right] \tag{10}$$

$$S = -\frac{D}{T} F \quad > 0$$

In Rindler space, $T = \alpha/2\pi$ and so

$$F = -\frac{\Gamma(D)\ \zeta(D)}{(4\pi)^{D-1/2}\ \pi^D} \, \frac{\alpha}{(D-2)} \, e^{\alpha H (D-2)} \tag{11}$$

$$S = \frac{\Gamma(D)\ \zeta(D)}{(4\pi)^{D-1/2}\ \pi^D} \, \frac{2\pi D}{(D-2)} \, e^{\alpha H (D-2)} \tag{12}$$

We explicitly see that F and S need the ultra-violet cut-off H to be finite. This is equivalent to considering the following mapping defining the accelerated co-ordinates

$$\begin{aligned} x^1 - x^0 + \varepsilon &= e^{\alpha (X^1 - X^0)} \\ x^1 + x^0 + \varepsilon &= e^{\alpha (X^1 + X^0)} \end{aligned} \tag{13}$$

where $\varepsilon \simeq e^{-\alpha H}$, $-H < X^1 < +\infty$.

As it is known, the string action in a D-dimensional (generically curved) space-time is given by

$$S = \frac{1}{2\pi\alpha'} \int d\sigma d\tau \sqrt{g}\, g^{\alpha\beta} G_{AB}(X)\, \partial_\alpha X^A \partial_\beta X^B \tag{14}$$

$G_{AB}(X)$ and $g_{\alpha\beta}(\sigma,\tau)$ stand for the space-time and the world-sheet metrics respectively, namely

$$\begin{aligned} dS^2 &= G_{AB}(X)\, dX^A dX^B , & 0 \leq A,B \leq D-1 \\ ds^2 &= g_{\alpha\beta}(x)\, dx^\alpha dx^\beta & 0 \leq \alpha,\beta \leq 1 \end{aligned} \tag{15}$$

$(2\pi\alpha')^{-1}$ is the string tension (we will take here $\alpha' = 1$).

The equations of motion and the constraints are

$$\Box^2 X^A + \Gamma^A_{BC}(X)\, \partial_\alpha X^B \partial_\beta X^C g^{\alpha\beta} = 0 \tag{16}$$

$$T_{\alpha\beta} = G_{AB}\, \partial_\alpha X^A \partial_\beta X^B - \frac{1}{2} g_{\alpha\beta}\, \partial_\sigma X^A \partial^\sigma X^B = 0 \tag{17}$$

$\Gamma^A_{BC}(X)$ stand for the Christoffel connections of the metric G_{AB}. As is well known, the action (14) is invariant under reparametrizations of the world-sheet co-ordinates

$$\delta X^\mu = \varepsilon^\alpha(\xi)\, \partial_\alpha X^\mu(\xi) , \qquad \xi^\alpha = (\sigma, \tau)$$

under which the two-dimensional metric $g_{\alpha\beta}$ goes to

$$\delta g_{\alpha\beta}(\xi) = \varepsilon^\gamma \partial_\gamma g_{\alpha\beta} + (\partial_\alpha \varepsilon^\gamma)\, g_{\gamma\beta} + (\partial_\beta \varepsilon^\gamma)\, g_{\alpha\gamma} .$$

This reparametrization invariance allows us to choose $g_{\alpha\beta}$ in the conformal form (so called "conformal gauge"), namely,

$$\begin{aligned} g_{\alpha\beta} &= \Lambda(\xi)\, \eta_{\alpha\beta} , \\ ds^2 &= \Lambda(\sigma,\tau)\,(d\sigma^2 - d\tau^2) \end{aligned} \tag{18}$$

The choice of the conformal gauge still allows the reparametrizations

$$x_+ + \varepsilon = f(x'_+ + \ell)$$
$$x_- + \varepsilon = g(x'_- + \ell) \qquad (19)$$

(where $x_\pm = \sigma \pm \tau$, $x'_\pm = \sigma' \pm \tau'$ and we have introduced the constants ε, ℓ whose meaning will be clear in the sequel). In this way, we can choose the light-cone gauge:

$$V \equiv X_0 + X_1 = p_+ x_+ \qquad (20)$$

where the proportionality constant $p_+ > 0$ is the momentum of the centre-of-mass of the string.

An important step in field as well as string quantization is the definition of positive frequency states and its associated ground state. It is by now well known, in QFT (in either flat or curved space-times) that "canonical states" for different co-ordinate systems are physically different (each timelike vector field leads to a separate indication of what constitutes a positive frequency)[1),2)]. There are different ways (and there is thus an ambiguity) in choosing such a basis. This makes it possible for a given field theory to have different alternative well-defined Fock spaces (different "sectors" of the theory). To illustrate this feature in string theory, let us consider the simplest case $G_{AB} = \eta_{AB}$, i.e.,

$$dS^2 = -dX^{0^2} + dX^{1^2} + \ldots\ldots + dX^{D-1^2}, \qquad (21)$$

which means both that the space-time is flat and that the string is described in an inertial frame. Usually the string is described in this frame where

$$\partial_{x_+} \partial_{x_-} X^A = 0 \quad , \qquad A = 0, 1, \ldots, D-1 \qquad (22)$$

and then positive frequency modes are defined with respect to the inertial time X^0. In the light-cone gauge, X^0 is proportional to τ and therefore the modes

$$\overleftarrow{\varphi}_n = \frac{1}{2\sqrt{\pi|n|}}\, e^{-inx_+} \quad , \quad \overrightarrow{\varphi}_n = \frac{1}{2\sqrt{\pi|n|}}\, e^{inx_-} \qquad (23)$$

define the (inertial) particle states of the string.

Obviously, if we consider the (σ',τ') parametrization of the string world sheet [Eq. (19)], i.e.,

$$\sigma' + \ell = \frac{1}{2}\left[G(\sigma+\tau+\varepsilon) + F(\sigma-\tau+\varepsilon)\right] , \quad F \equiv f^{-1}$$

$$\tau' = \frac{1}{2}\left[G(\sigma+\tau+\varepsilon) - F(\sigma-\tau+\varepsilon)\right] , \quad G \equiv g^{-1} \tag{24}$$

for which we have

$$\partial_{x'_+} \partial_{x'_-} X^A = 0 , \tag{25}$$

positive frequency modes with respect to τ', namely

$$\overleftarrow{\phi}_n = \frac{1}{2\sqrt{\pi |n|}} e^{-i\lambda_n x'_+} , \qquad \overrightarrow{\phi}_n = \frac{1}{2\sqrt{\pi |n|}} e^{i\lambda_n x'_-} \tag{26}$$

do not define positive frequency modes with respect to X^0, i.e., are not inertial particle states of the string. However, they are positive frequency modes with respect to another (accelerated) time $X^{0'}$ defined by

$$X_1 - X_0 = p_- f\,(X'_1 - X'_0)$$
$$X_1 + X_0 = p_+ g\,(X'_1 + X'_0) \tag{27}$$

where f is the same as in Eq. (19). This corresponds to the description of the string in an accelerated reference frame $\{X'_0, X'_1, \ldots, X'_{D-1}\}$, with acceleration

$$\tilde{a} = \frac{1}{[f'g']^{1/2}} \partial_{X'_1} \left[\ln (f'g')\right]$$

and metric

$$dS^2 = p_+ p_- f'(X'_1 - X'_0)\, g'(X'_1 + X'_0)\left[-dX'^2_0 + dX'^2_1\right] + dX'^2_2 + \ldots + dX'^2_{D-1} \tag{28}$$

where, for simplicity, we have taken

$$X_i' = X_i \quad , \quad (i = 2, \ldots, D-1)$$

For the description of the string in this frame we can always choose

$$V \equiv X'_1 + X'_0 = x'_+ + \ell \tag{29}$$

as it follows from Eqs. (19), (20) and (27). Thus, Eq. (25) defines the accelerated particle states of the string. [The choice (29) is particularly useful here because of the acceleration points in the X^1 direction]. In this frame, the equations of motion of the string are

$$\partial_{x'_+}\partial_{x'_-} U' + \frac{f''}{f'}\partial_{x'_+}U'\partial_{x'_-}U' + \left(\frac{f''}{f'} - \frac{g''}{g'}\right)\partial_{x'_+}U'\partial_{x'_-}V' = 0$$

$$\partial_{x'_+}\partial_{x'_-} V' + \frac{g''}{g'}\partial_{x'_+}V'\partial_{x'_-}V' - \left(\frac{f''}{f'} - \frac{g''}{g'}\right)\partial_{x'_+}U'\partial_{x'_-}V' = 0 \tag{30}$$

$$\partial_{x'_+}\partial_{x'_-} X'^i = 0 \tag{31}$$

The equations of motion of the physical (transverse degrees) of freedom are the same in both frames. The equations of motion of the longitudinal co-ordinates (U',V') are different, but it can be shown[3)] that, as in the inertial case[4)], they can be eliminated in terms of the transverse co-ordinates by using the constraints

$$T'_{\pm\pm} = f'(U')\, g'(V')\, \partial_{\pm'} U\, \partial_{\pm'} V + \partial_{\pm'} X^i \partial_{\pm'} X^i \approx 0$$

$$T'_{+-} = 0 \tag{32}$$

The Poincaré invariance of flat space-time has a non-linear realization in terms of the accelerated co-ordinates, but an explicit realization of the Poincaré algebra can be constructed in terms of the accelerated modes provided (at the quantum level) one sets D = 26 and uses symmetric ordering of the operators[3)].

The manifold (σ',τ') defined by Eq. (24) is the convenient world sheet parametrization for an accelerated string or a string in an accelerated space-time. This is true in either flat or curved space-time [although in the last case, Eq. (25) can only be valid asymptotically for $x'_\pm \to \pm\infty$].

Let us describe the string in the reference frames (21) and (28) referred in the sequel as (I) and (A) respectively. We consider here a closed string and take for simplicity f ≡ g. The boundary conditions are

$$X^{\mu}(0,\tau) = X^{\mu}(\pi,\tau)$$
$$X^{\mu}(0,\tau') = X^{\mu}(L_{\varepsilon},\tau') \tag{33}$$

Notice that

$$0<\sigma<\pi \qquad\qquad -\infty < X^{\circ} < +\infty \; .$$
$$-\infty<\tau<+\infty \qquad , \tag{34}$$

In order to have

$$0<\sigma'<L_{\varepsilon} \qquad\qquad -\infty < X'^{0} < +\infty \; ,$$
$$-\infty<\tau'<+\infty \qquad , \tag{35}$$

appropriate boundary conditions on the mappings must be imposed. Condition (33) is automatically satisfied from Eqs. (19) with

$$L_{\varepsilon} = F(\varepsilon+2\pi) - F(\varepsilon) \qquad\qquad F\equiv f^{-1}$$
$$\ell = F(\varepsilon) \qquad , \tag{36}$$

Condition (35) on τ' and X'° is not particular to strings. It is required to get a consistent quantization (of fields or strings) in accelerated manifolds and to have a complete in (out) basis[2)]. On the other hand, the manifold A can cover only a (bounded) region (namely $|X^1| > |X^{\circ}|$) of the original (I) one. Similarly, the manifold (σ',τ') can cover only a domain ($|\sigma| > \tau$) of the inertial or global world sheet. All these considerations are satisfied by taking mappings such that

$$u_{\pm} = f(\pm\infty) \; , \tag{37}$$

where $u_{+}(u_{-})$ are constants which can take independently finite or infinite values and $u_{+} > u_{-}$. This means that the inverse mapping $F \equiv f^{-1}$ has singularities at $u = u_{\pm}$, i.e.,

$$F(u_{\pm}) = \pm\infty \tag{38}$$

Singularities of these mappings describe the asymptotic regions of the space-time. Critical points of f, i.e.,

$$f'(\pm\infty) = 0 \tag{39}$$

describe event horizons at $u = u_-$ and $u = u_+$. In this case $u_-(u_+)$ are finite and the manifold A cover the region $u_- < |X^1 \pm X^0| < u_-$ of the global (I) space-time. If $u_\pm = \pm\infty$, there are no horizons. In this approach, the well-known Rindler's space corresponds to the mapping

$$f = e^{\alpha U'} \qquad , \quad \alpha = \text{const.} \tag{40}$$

In this case, $u_- = 0$ and $u_+ = +\infty$; $u = u_- = 0$ is an event horizon and the manifold covers the right-hand wedge $|U| > 0$. (In order to cover the whole plane, four Rindler's patches are needed.) For the string world sheet we have

$$x_\pm + \varepsilon = e^{\alpha (x'_\pm + \ell)} \tag{41}$$

and we see how the presence of the constant ε is necessary in order to have a finite string period in the manifold (σ', τ'). For the Rindler's mapping (41) we have

$$\ell = \frac{1}{\alpha} \log \varepsilon$$

$$L_\varepsilon = \frac{1}{\alpha} \log \left(\frac{2\pi}{\varepsilon} + 1 \right) \tag{42}$$

For $\varepsilon \to 0$, there is a stretching effect of the string due to the presence of an event horizon in the world sheet.

Let us now consider the string quantization in the frames (I) and (A). The string co-ordinates can be split into left and right movers as

$$X^A(\xi) = \overleftarrow{X}^A(\xi_+) + \overrightarrow{X}^A(\xi_-) \tag{43}$$

where $\xi_\pm$ stands for $\sigma \pm \tau$ or $\sigma' \pm \tau'$. As usual, left and right modes of the closed string decouple both in the classical and quantum theories. We consider first the left movers. The wave modes are periodic functions of x_+ and x_- with period 2π. The left modes which are only functions of x_+ will be periodic functions of $x_+' = \sigma' + \tau'$ with period L_ε. In the inertial frame (I) we have

$$\overleftarrow{X}^i(x_+) = \frac{q^i}{2} + \frac{p^i}{2} x_+ + \frac{1}{2} \sum_{n \neq 0} \frac{\overleftarrow{a}_n^i}{\sqrt{|n|}} e^{-inx_+}$$

$$\overleftarrow{U}(x_+) = q_- + p_- x_+ + \frac{i}{2} \sum_{n \neq 0} \frac{\overleftarrow{\mu}_n}{n} e^{-inx_+} \qquad p_- = \frac{\mu_0}{2} \tag{44}$$

The independent dynamical variables $\overleftarrow{a}_n^i (n \in z)$, q_i, q_+ and p_+ obey the canonical commutation relations

$$[\overleftarrow{\alpha}_n^i , \overleftarrow{\alpha}_m^j] = n\, \delta_{n+m}\, \delta^{ij} , \quad \overleftarrow{\alpha}_n^i = -i\sqrt{|n|}\, \mathrm{sgn}\, \overleftarrow{a}_n^i , \ n \neq 0$$

$$[q^i , \overleftarrow{\alpha}_0^j] = i\, \delta^{ij} , \quad \overleftarrow{\alpha}_0^i = p^i \tag{45}$$

$$[q_- , p_-] = i$$

Then

$$[\overleftarrow{X}^i(x_+), \overleftarrow{p}^i(y_+)] = \frac{\delta^{ij}}{2} \delta(x_+ - y_+)$$

The energy-momentum tensor, the conformal generators and the mass operator are

$$T_{++}(x_+) = \sum_{n \in z} \frac{L_n}{2\pi} e^{-inx_+} ,$$

$$L_n = \frac{1}{2} \sum_m \overleftarrow{\alpha}_m^i \overleftarrow{\alpha}_{n-m}^i + p_+ \overleftarrow{\mu}_n \tag{46}$$

$$L_0 = \frac{1}{2} \sum_m \overleftarrow{\alpha}_m^i \overleftarrow{\alpha}_{-m}^i + 2 p_+ p_-$$

and

$$M^2 = p_+ p_- - p_i^2 = \sum_m \frac{n}{2} (a_n^{i+} a_n^i + a_n^i a_n^{i+})$$

$$= \sum_m n\, a_m^{i+} a_m^i - \frac{(D-2)}{24} \tag{47}$$

In the accelerated frame (A) we have

$$\overleftarrow{X}^i(x'_+) = \frac{Q^i}{2} + \frac{P^i}{2}x'_+ + \sum_{n\neq 0} \overleftarrow{C}_n^{\,i}\, \overleftarrow{\phi}_n(x'_+)$$

$$\overleftarrow{U}(x'_+) = u_0 + \frac{P_-}{2}x'_+ + \frac{i}{2}\sum_{n\neq 0}\frac{\gamma_n}{n} e^{-i\lambda_n x'_+} \tag{48}$$

with

$$\overleftarrow{X}(x'_+ + L_\varepsilon) - \overleftarrow{X}(x'_+) = \frac{P^i}{2\pi} L_\varepsilon$$

$$\overleftarrow{U}(x'_+ + L_\varepsilon) - \overleftarrow{U}(x'_+) = \frac{P_-}{2} L_\varepsilon \tag{49}$$

and

$$P^i = \frac{2\pi}{L_\varepsilon} p_i \quad , \quad P_- = \frac{2\pi}{L_\varepsilon} p_-$$

The left modes provide the T'_{++} component of the energy-momentum tensor. The accelerated conformal generators (L'_n) follow from the Fourier transform of T'_{++}:

$$T'_{++}(x'_+) = \sum_{-\infty}^{\infty} \frac{L'_n}{L_\varepsilon} e^{-i\lambda_n x'_+}$$

$$L'_n = L'^{\perp}_n + L'^{\parallel}_n \simeq 0 \tag{50}$$

$$L'^{\perp}_n = -\frac{1}{L_\varepsilon}\sum_m \overleftarrow{\beta}_m^{\,i}\, \overleftarrow{\beta}_{n-m}^{\,i}$$

$$\overleftarrow{C}_n^{\,i\,+} = i\,\mathrm{sgn}\, \overleftarrow{\beta}_n^{\,i}/\sqrt{|n|} \; , \; \overleftarrow{\beta}_n^{\,i} = \overleftarrow{\beta}_{-n}^{\,i} \; , \; \beta_0 = p^i \; , \; \gamma_0 = p_- .$$

We find the following Fock representation

$$[\overleftarrow{C}_n^{\,i}, \overleftarrow{C}_m^{\,j}] = \delta_{nm}\,\delta^{ij} \; , \; [\overleftarrow{\beta}_n^{\,i}, \overleftarrow{\beta}_m^{\,j}] = n\,\delta_{n+m,0}\,\delta^{ij}$$

$$[P^i, Q_0^j] = -\left(\frac{2\pi}{L_\varepsilon}\right) i\,\delta^{ij} \tag{51}$$

$$[p_+, u_0] = -i$$

which implies

$$[\overleftarrow{X}^i(\xi), \overleftarrow{P}^j(\xi')] = \frac{i}{2}\,\delta(\xi - \xi')\,\delta^{ij} \tag{52}$$

For a string whose centre-of-mass follows an accelerated world line in the longitudinal direction X^1, of the type

$$x^1 \pm x^0 \pm a_\pm = A_\pm \, e^{\pm\alpha\tau'} \,, \quad a_\pm, A_\pm = \text{const.},$$

it is convenient to use the expansion[3)]

$$U = U_0 + \frac{1}{f(x'_+)} \left[\tilde{p}_- + \frac{i}{2} \sum_{n\neq 0} r_n \, e^{-i\lambda_n x'_+} \right] \qquad (53)$$

for which the mass formula is given by

$$M'^2 = (\alpha L_\varepsilon)^2 p_+ \tilde{p}_- - p_i^2 = \sum_1 \frac{n}{2} \left(C_n^{i+} C_n^i + C_n^i C_n^{i+} \right) =$$
$$= \sum_1 n \, C_n^{i+} C_n^i - \frac{(D-2)}{24} \qquad (54)$$

The operators M^2 and M'^2 [Eqs. (47) and (54)] are different but both have the same eigenvalues.

The same treatment applies to the right movers $\vec{X}^A$ in a gauge defined by

$$\vec{U}' = p_- x_-$$

and with $\vec{U}' = x'_- + l$,

i.e. $x_- + \varepsilon = f(x'_- + l)$

The (I) and (A) Fock representations are related by a Bogoliubov transformation

$$C_m^i = p^i O_m + \sum_{n=1}^{\infty} \left(A_{mn} a_n^i + B_{mn} a_n^{i+} \right), \quad m \neq 0 \qquad (55)$$

with Bogoliubov coefficients

$$A_{mn} = \langle \phi_{\lambda m}, \varphi_n \rangle = \frac{1}{2\pi} \sqrt{\frac{n}{m}} \int_0^{2\pi} e^{in\sigma - i\lambda_m F(\sigma+\varepsilon)} \, d\sigma$$

$$B_{mn} = \langle \phi_{\lambda m}, \varphi_n^* \rangle = \frac{1}{2\pi} \sqrt{\frac{n}{m}} \int_0^{2\pi} e^{in\sigma + i\lambda_m F(\sigma+\varepsilon)} \, d\sigma \qquad (56)$$

and zero-mode contribution

$$O_m = \left\langle \frac{1}{2} \left(x_+ - \frac{2\pi}{L_\varepsilon} x'_+ \right), \phi_{\lambda m} \right\rangle$$

The ground state of the string is defined by

$$\begin{aligned} a_n^i\,|0\rangle &= 0 \qquad \forall n > 1 \\ p^i\,|0\rangle &= 0 \end{aligned} \tag{57}$$

We also have

$$\begin{aligned} C_m\,|0'\rangle &= 0 \qquad \forall m > 1 \\ p^i\,|0'\rangle &= 0 \end{aligned} \tag{58}$$

but $C_m|0\rangle \neq 0$. The expectation value of the accelerated number operator in the ground state $|0\rangle$ is equal to (not sum over i)

$$N^i(\lambda_m) \equiv \langle 0|\, C_m^i\, C_m^{i\,+}\,|0\rangle = N(\lambda_m) = \sum_{n=1}^{\infty} |B_{mn}|^2 \tag{59}$$

which is i-independent. The expectation value of the accelerated M'^2 operator is equal to

$$M'^2 = 24 \sum_{1}^{\infty} n\, N(n) - 1 \tag{60}$$

where N(n) is given by Eq. (59) with $\lambda_n = (2\pi/L_\varepsilon)n$. If $f(x'_+ + \ell) = e^{\alpha(x'_+ + \ell)}$, then:

$$\begin{aligned} B_{mn}\,(n \gg 1) &= \mathcal{C}(\lambda_m)\, \frac{n^{-i\lambda_m/\alpha}}{\sqrt{n}} + \frac{b}{\sqrt{n}} + O\left(\frac{1}{n^{3/2}}\right) \\ \mathcal{C} &= \frac{i}{2\pi}\, \frac{e^{-\frac{\pi\lambda_m}{2\alpha}}}{\sqrt{m}}\, \Gamma\left(1 + \frac{i\lambda_m}{\alpha}\right), \\ b &= -\frac{i}{2\pi} \left(\frac{2\pi}{\sqrt{m}}\right)^{i\lambda_m/\alpha} \end{aligned} \tag{61}$$

and

$$\begin{aligned} N(\lambda_m) &= \frac{1}{\left(e^{2\pi\lambda_m/\alpha} - 1\right)} + \frac{\alpha}{2\pi\lambda_m} \\ M'^2 &= 24\left[\sum_{n=1}^{\infty} \frac{n}{\left(e^{n/T_0} - 1\right)} + T_0\right] - 1 \end{aligned} \tag{62}$$

$$M'^2_{\varepsilon \ll 1} = \frac{1}{2\pi} \ln \frac{2\pi}{\varepsilon} \tag{63}$$

[Recall that in field theory, $B_{\lambda k}\,(k \gg 1) = \mathcal{C}(\lambda)\,(k/\sqrt{k})^{-i\lambda/\alpha}$, with $\lambda \in R$.] $N(\lambda_n)$ [Eq. (62)] is equal to a Planckian spectrum at the Hawking-Unruh temperature

$$T_s = \frac{\alpha}{2\pi} \qquad (64)$$

In N(n), the temperature is

$$T_0 = \left(\frac{L_\varepsilon}{2\pi}\right) T_s = \frac{a}{2\pi} \underset{\varepsilon \ll 1}{=} \frac{1}{(2\pi)^2} \ln\left(\frac{2\pi}{\varepsilon}\right) \qquad (65)$$

The expectation value of M'^2 in the ground state $|0\rangle$ is positive. The tachyon level of this state is filled up with accelerated modes following a thermal distribution. For any other mapping f different from the exponential one [Eq. (41)], B_{nm} $(n \gg 1) \to e^{-n}$ and $N(\lambda_n)$ and N(n) are non-thermal.

It can be noted that the acceleration as well as the temperature T_s appearing in N(n) are re-scaled by the factor $(L_\varepsilon/2\pi)$, i.e., the ratio of the string period in the inertial and in the accelerated frame. Thus, the dimensionless temperature T_0 does not depend on the acceleration parameter α of Rindler's space, but on the parameter

$$a = \left(\frac{L_\varepsilon}{2\pi}\right) \alpha = \frac{1}{2\pi} \ln\left(\frac{2\pi}{\varepsilon} + 1\right) \underset{\varepsilon \ll 1}{=} -\frac{\ell_d}{2\pi} \qquad (66)$$

We see that the parameter ε $[\ell \equiv \ell(\varepsilon)]$ introduced in the mappings on the world sheet, plays a fundamental rôle in the string context. ε acts as a regulator or cut-off to avoid the presence of an event horizon in the world sheet and to get a finite string period. Its magnitude is of the order of the Planck length.

TRANSFORMATION BETWEEN THE STATES. DERIVATION OF THE PARAMETER ε FROM THE CONSTRAINT EQUATIONS

We see that the transformations (19) are not without consequences but change the ground state and in general, the quantum states, except for mappings belonging to the O(2,1) group. Such invariance mappings are the Möbius or bilinear transformations. Under the transformations (19) and (27), the states transform as

$$|\,\rangle \to |\,'\rangle = e^{i\hat{G}}\,|\,\rangle \qquad (67)$$

where $a_n|\,\rangle = 0$, $C_n|\,'\rangle = 0$, $\forall_n$ and

$$\hat{G} = \sum_n \theta_n \left(C_n C_n - C_n^+ C_n^+\right)$$

[this is an operatorial representation for the Bogoliubov transformation, Eq. (55), with Bogoliubov coefficients $\cosh\theta_n$ and $\sinh\theta_n$].

The vacuum expectation value of $T_{\mu\nu}$ transforms as

$$\langle T_{\mu\nu} \rangle = \langle T_{\mu\nu}{}' \rangle + \Theta_{\mu\nu} + P_{\mu\nu} \tag{68}$$

$P_{\mu\nu}$ is any conserved traceless tensor taking into account the dependence of $\langle T_{\mu\nu} \rangle$ on the quantum state. It represents the non-local part of $T_{\mu\nu}$. $\Theta_{\mu\nu}$ depends on the mapping and represents the local part. In the conformal gauge we have

$$P_{--} = \mathcal{U}_-(x'_-) \quad , \quad P_{++} = \mathcal{U}_+(x'_+) \; , \; P_{+-} = P_{-+} = 0 \tag{69}$$

$$\Theta_{--} = -\frac{1}{12\pi} \sqrt{f'}\, d^2_{x'_-} \left(\frac{1}{\sqrt{f'}} \right)$$

$$\Theta_{++} = -\frac{1}{12\pi} \sqrt{g'}\, d^2_{x'_+} \left(\frac{1}{\sqrt{g'}} \right) \tag{70}$$

$$\Theta_{+-} = \Theta_{-+} = 0$$

$\mathcal{U}_-$ and $\mathcal{U}_+$ are arbitrary functions of the indicated variables. Equations (69) derive from the conditions $\nabla^\nu P_{\mu\nu} = 0$ and $P_\nu{}^\nu = 0$. Therefore, the constraints

$$\langle T_{\mu\nu} \rangle = 0 \quad , \quad \langle T_{\mu\nu}{}' \rangle = 0$$

yield to the equations[5)]

$$\sqrt{f'}\, d^2_{x'_-} \left(\frac{1}{\sqrt{f'}} \right) - 12\pi\, \mathcal{U}_-(x'_-) = 0 \tag{71}$$

$$\sqrt{g'}\, d^2_{x'_+} \left(\frac{1}{\sqrt{g'}} \right) - 12\pi\, \mathcal{U}_+(x'_+) = 0$$

These are zero-energy Schrödinger equations

$$\frac{d^2}{dx'^2_\pm} \psi_\pm - 12\pi\, \mathcal{U}_\pm(x'_\pm)\, \psi_\pm = 0 \tag{72}$$

for the mappings f and g. By giving the potentials $\mathcal{U}_\pm$, Eqs. (72) determine the wave functions

$$\Psi_- = \frac{1}{\sqrt{f'}} \quad , \quad \Psi_+ = \frac{1}{\sqrt{g'}} \tag{73}$$

Because $\mathcal{U}_\pm$ are arbitrary functions [compatible with the boundary conditions (37)], Eqs. (72) do not yield additional constraints on the mappings f and g, but a way of connecting the mapping to a potential problem.

The first term of Eq. (71) is the Schwarzian derivative of f:

$$D[f] = \frac{f'''}{f'} - \frac{3}{2}\left(\frac{f''}{f'}\right)^2 \quad , \tag{74}$$

which is invariant under the Möbius or bilinear transformations. Under these transformations, f becomes a new function but D[f] is invariant determining the same ground state of the string. In particular, $\mathcal{U}_+ = \mathcal{U}_- = 0$ determine f(g) as

$$f = \frac{\alpha x'_- + \beta}{\gamma x'_- + \delta} \quad , \quad (\alpha\delta - \beta\gamma) = 1 \tag{75}$$

The ground state defined by this mapping can be considered as a reference or "minimal" state at zero temperature with respect to which other ground states corresponding to non-zero potentials $\mathcal{U}_\pm$ appear as excited or thermal ones. If $\mathcal{U}_+ = \mathcal{U}_- = \mathcal{U}_0$ = constant > 0, then

$$\Psi_\pm = A\, e^{-K x'_\pm} \quad , \quad f = \frac{1}{\alpha A^2}\, e^{\alpha x'_+} - \varepsilon \tag{76}$$

where A is a normalizing constant (we choose $A^2 = e^{-\alpha\ell}/\alpha$) and K is the zero-energy transmission coefficient

$$K \equiv \frac{\alpha}{2} = \sqrt{12\pi\,\mathcal{U}_0} \quad ,$$

For $\varepsilon \to 0$, the mapping (76) defines an event horizon at $x_\pm = 0$ ($x'_\pm = -\infty$) and carries an intrinsic temperature $T_s = \alpha/2\pi$, as can be seen by putting $t = i\tau$ ($x_+ = x+i\tau$) and then $0 < \tau' < 2\pi/\alpha$. The temperature appears related to the height of the potential, namely

$$T_s = \sqrt{\frac{12\,\mathcal{U}_0}{\pi}}$$

and the parameter ε arises naturally as an integration constant. This temperature T_s characterizes the spectrum $N(\lambda_n)$, Eq. (62).

Similar equations to (71) also appear in the so-called "back reaction problem" in two dimensions[6)] [as a consequence of the (±±) components of the semi-classical Einstein equations, the (+-) component giving rise to the Liouville equation for the geometry because of the conformal anomaly]. Equations (71) can also be derived in the context of conformal field theories on higher genus Riemann surfaces (the potential playing the rôle of the zero-point energy) in connection with the approach of Ref. 7).

We have shown that with appropriate boundary conditions, the holomorphic mappings of reparametrization invariance of string theory can be interpreted as a change of co-ordinate frame in the space-time in which strings are embedded. These mappings change the ground state in the quantized theory except for transformations belonging to the O(2,1) group. This allowed us to discuss in a systematic way the Hawking-Unruh effect in string theory. The transformations describing the world sheet of an accelerated string need the introduction of an additional parameter ε with respect to those describing the trajectories of accelerated point particles.

The results found here apply also to curved space-time. For the most important metrics in general relativity, the presence of isometry groups allows the maximal analytic extension of the (D-dimensional) manifold to be performed through the extension of a relevant two-dimensional manifold containing the time axis and a suitable spatial co-ordinate. This maximal analytic extension is performed by mappings like [Eq. (27)], where X_1, X_0 are Kruskal (maximal) type co-ordinates and X_1', X_0' are of the Schwarzschild type. [For the role played by these mappings in the context of Q.F.T., see Refs. 8) and 9).]

STRINGS NEAR BLACK HOLES

Our investigation of strings in Rindler space-time can be applied to the case of strings in a black hole background. Black hole solutions of Einstein equations exist in D-space-time dimensions (D > 4)[10)]. These solutions are asymptotically flat and generalize the Schwarzschild space-time of four dimensions; they have the metric

$$ds^2 = -\left(1 - \frac{C}{R^{D-3}}\right) dT^2 + \frac{dR^2}{\left(1 - \frac{C}{R^{D-3}}\right)} + R^2 d\Omega^2_{D-2} \qquad (77)$$

R is the radial co-ordinate, $d\Omega^2_D$ is the line element on the unit D-sphere and the constant C is > 0. The surface

$$R = C^{1/D-3} \equiv R_S$$

is an event horizon (there are both past and future event horizons) and R = 0 is a space-like singularity. The horizon radius R_S is related to the black hole mass M by

$$C = 16\pi G \frac{M}{(D-2)\, A_{D-2}}$$

where

$$A_{D-2} = \frac{2\pi^{(D-1)/2}}{\Gamma\left(\frac{D-1}{2}\right)}$$

is the area of a unit (D-2) sphere and G has dimensions of length^{D-2}.

The mass and the surface gravity K of the black hole are related by

$$K = \left(\frac{D-3}{2 R_S}\right) = \frac{(D-3)}{2}\left[\frac{(D-2)\, A_{D-2}}{16\pi G M}\right]^{\frac{1}{D-3}} \qquad (78)$$

For D = 4 this yields the standard relations $R_S = 2GM$ and $K = 1/(4GM)$.

The Kruskal extension of this Schwarzschild manifold is given by the mapping

$$r_K \pm t_K = e^{K(R^* \pm T)} \qquad (79)$$

where

$$R^* = R + R_S^{D-3} \int \frac{dR}{(R^{D-3} - R_S^{D-3})}$$

$$-\infty \leq R^*, T \leq +\infty \, .$$

This is the same exponential mapping as the Rindler transform with K instead of α. The Rindler co-ordinates are similar to the Schwarzschild (R^*, T) ones and the Minkowskian co-ordinates are similar to the Kruskal (global) co-ordinates (r_K, t_K). The event horizon $R = R_S$ corresponds to $R^* = -\infty$. As discussed above, a large cut-off ($H \gtrsim 1/K$) is needed in the negative Schwarzschild co-ordinate R^*. This shifting of the horizon is equivalent to considering a shifting ε in the mapping

$$r_K \pm t_K + \varepsilon = e^{K(R^* \pm T)} \tag{80}$$

with

$$\varepsilon = e^{-KH} \sim l_p^2$$

and thus

$$-H \leq R^* \leq +\infty$$

reflecting the fact that a classical description of the geometry is no longer valid at distances of order of the Planck length. We will take

$$\varepsilon = \frac{\lambda_c}{R_s} \quad \text{, where} \quad \lambda_c = \frac{1}{M}$$

is the Compton length of the black hole (here $\hbar = c = 1$). Thus the shifting of the horizon is $H = (1/K) \ln\varepsilon$, with

$$\varepsilon = \pi^{1/2} \left[\frac{(D-2)}{8\, \Gamma\left(\frac{D-1}{2}\right)} \right]^{\frac{1}{D-3}} \left(\frac{M_{Pl}}{M} \right)^{\frac{D-2}{D-3}} \tag{81}$$

and

$$K = \frac{\pi^{1/2}}{2} (D-3) \left[\frac{(D-2)}{8\, \Gamma\left(\frac{D-1}{2}\right)} \right] \left(\frac{M_{Pl}}{M} \right)^{\frac{1}{D-3}} M_{Pl} \tag{82}$$

Following on the same lines of argument discussed above, for the choice of gauge and parametrizations of the string world sheet and considering only left movers, we have

$$\overleftarrow{v} \equiv r_K + t_K = p_+ x_+ \tag{83}$$

$$V \equiv R^* + T = x'_+ + \frac{1}{K} \log p_+ \qquad (84)$$

and

$$x_+ + \varepsilon = e^{K x'_+}$$

(Here the longitudinal direction of the string is in the radial direction.)

Positive frequency modes, $\overleftarrow{\psi}_n$ with respect to the Kruskal time t_K and $\overleftarrow{\phi}_n$ with respect to the Schwarzschild time T can be defined. The Schwarzschild frequency is equal to

$$\lambda_n = \frac{2\pi}{L_\varepsilon} n \quad , \qquad n = 1, 2, \ldots \qquad (85)$$

where
$$L_\varepsilon = \frac{1}{K} \ln \left(\frac{2\pi}{\varepsilon} + 1 \right)$$

For $\varepsilon \ll 1$, the frequency spacing tends to zero reflecting the stretching effect of the string near the horizon as seen by a Schwarzschild external observer. Associated to the modes $\overleftarrow{\psi}_n$ and $\overleftarrow{\phi}_n$ we will have Kruskal and Schwarzschild operators a_n and C_n respectively and a vacuum state defined by

$$a_n |0_K\rangle = 0 \quad , \qquad \forall n > 1$$

On the other hand, in order to have a smooth Euclidean manifold from a black hole space-time with topology $R^2 \times S^{D-2}$, the Schwarzschild imaginary time iT must be identified with a period

$$\beta = \frac{2\pi}{K}$$

Then the same periodicity in the imaginary time appears in the correlation functions of string co-ordinates, indicating that the string is in equilibrium with a heat bath at the Hawking temperature

$$T_s = \frac{K}{2\pi} \qquad (86)$$

The same temperature T is recovered in the function $N(\lambda_m)$, i.e.,

$$N(\lambda_n) = \langle 0_K | C_{\lambda_n}^{+} C_{\lambda_n} | 0_K \rangle ,$$

which gives a Planckian distribution for the Schwarzschild modes but with a "filter" $|g(\lambda_n)|^2$ equal to the absorption cross-section of the black hole. In the spectrum N(n) in which frequency is measured in dimensionless units 1,2,..., the temperature of the Planckian distribution is equal to

$$T_0 = \frac{1}{4\pi^2} \ln\left(\frac{2\pi}{\varepsilon}\right) \gg 1 \qquad (87)$$

One can consider different higher dimensional black hole space-times, namely a 26-dimensional or a four-dimensional black hole with the extra 22 dimensions compactified in a torus[11)]. Intermediary situations can also be envisaged but it must be noticed that the qualitative properties of the string quantization will be the same since they depend upon the horizon structure in the two variables R, T (or X^0, X^1, for Rindler space). We hope to come back to this problem elsewhere.

It can be noticed that the Hagedorn temperature (T_m) in this context is

$$T_m = \frac{1}{\sqrt{\alpha'}} \sim M_{pl}$$

Then,

$$\frac{T_s}{T_m} \sim \left(\frac{M_{pl}}{M}\right)^{\frac{1}{D-3}}$$

and we have

$$T_s \lesssim T_m$$

since the basic requirement of the present semiclassical treatment is $M \gtrsim M_{pl}$.

NEW APPROACH TO STRING QUANTIZATION IN CURVED SPACE-TIMES

The main feature of strings propagating in a curved space-time is that the equations of motion [Eq. (16)] are non-linear in X_A, so right and left movers interact with each other and also with themselves. It must be noticed that purely left modes (or right modes) are exact solutions of Eq. (16), namely

$$X_A = q_A(\sigma + \tau) \qquad \text{or} \qquad X_A = q_A(\sigma - \tau) \tag{88}$$

When G_{AB} is the metric of a symmetric space, the equations of motion possess an associated linear system and exact solutions can be constructed, by using the inverse scattering method. In Ref. 12), we propose a general perturbative scheme to solve the equations of motion and constraints both classically and quantum mechanically. We start from an exact given solution of Eqs. (16) and develop in perturbations around. A possible starting point is a solution for the centre-of-mass motion of the string $q_A(\tau)$ where

$$\ddot{q}^A(\tau) + \Gamma^A_{BC}(q)\, \dot{q}^B(\tau)\, \dot{q}^C(\tau) = 0 \tag{89}$$

The world-sheet time variable τ is identified here with the proper time of the centre-of-mass trajectory. Another possibility is to take pure left(right) mover solutions. Then we set

$$X_A(\sigma,\tau) = q_A(\sigma,\tau) + \eta_A(\sigma,\tau) + \xi_A(\sigma,\tau) + \cdots \tag{90}$$

Here $q_A(\sigma,\tau)$ is an exact solution of Eq. (16) and η_A is a solution of the linearized perturbation around q_A. That is

$$\partial^2\eta^A + \Gamma^A_{BC}(q)\left(\partial_- q^B\partial_+ + \partial_+ q^B \partial_-\right)\eta^C + \partial_C \Gamma^A_{BD}(q)\, \partial_+ q^B\, \partial_- q^D\, \eta^C = 0 \tag{91}$$

$\xi_A(\sigma,\tau)$ fulfil the second order perturbation equation around q_A:

$$\begin{aligned}\partial^2\xi^A &+ \Gamma^A_{BC}\, \partial_+\eta^B\, \partial_-\eta^C + \frac{1}{2}\eta^D\eta^E\left(\partial^2_{DE}\Gamma^A_{BC}\right)\partial_+ q^B\, \partial_- q^C + \\ &+ \eta^D\left(\partial_D \Gamma^A_{BC}\right)\left(\partial_+ q^B\partial_- + \partial_- q^B \partial_+\right)\eta^C + \\ &+ \Gamma^A_{BC}\left(\partial_+ q^B \partial_- + \partial_- q^B\partial_+\right)\xi^C + \left(\partial_D\Gamma^A_{BC}\right)\partial_+ q^B\, \partial_- q^C\, \xi^D = 0\end{aligned} \tag{92}$$

One can consider the higher-order perturbations but in this note we will restrict ourselves to first and second orders.

It must be noticed that we are treating the space-time metric exactly and taking the string oscillations around its centre-of-mass $q_A(\tau)$ as perturbation. So, our expansion corresponds to low excitations of the string as compared with the energy scales of the metric G_{AB}. For example, this method is exact in flat space-time. In the Schwarzschild geometry, it will correspond to an expansion in ω/M where ω is the frequency mode and M the black hole mass. Thus, our approximation applies to black hole masses larger than the string energy. In other words, this corresponds to a strong gravitational field expansion. This can be equivalently considered as an expansion in powers of $\sqrt{\alpha'}$. Actually, since $\alpha' = \ell_{p\ell}^2$, ($\ell_{p\ell}$ is the Planck length), the expansion parameter turns out to be the dimensionless constant

$$g \equiv \frac{\sqrt{\pi}\, l_p}{R_c} = \frac{1}{l_p M} \simeq \frac{\omega}{M},$$

where the length R_c characterizes the curvature radius of the space-time under consideration and M its associated mass (the black hole mass in the Schwarzschild geometry, the mass of the Universe in a cosmological model). In most of the interesting situations, one clearly has $g \ll 1$.

It must be noticed that even for small g, the metric and its derivatives may be very large in some regions of the space-time. This shows that our method has a larger domain of applicability than the background field method where one must have everywhere

$$\left| \sqrt{\alpha'}\, \partial_c G_{AB}(x) \right| \ll 1$$

The first order equation describes the interaction between the string modes and the curved space-time geometry. The interactions between the string modes themselves start to appear from the second order perturbation (ξ^A) equation.

The constraint equations must also be expanded in perturbations. We find up to terms higher than the second order:

$$T_{\pm\pm} = G_{AB}(q)\, \partial_\pm q^A \partial_\pm q^B + 2\, G_{AB}\, \partial_\pm q^A \partial_\pm \eta^B +$$

$$+\eta^C \partial_C G_{AB}(q)\, \partial_\pm q^A \partial_\pm q^B + \frac{1}{2}\eta^C\eta^D \partial_{CD} G_{AB}(q)\, \partial_\pm q^A \partial_\pm q^B +$$

$$+ 2\eta^C \partial_C G_{AB}(q)\, \partial_\pm q^A \partial_\pm \eta^B + \xi^C \partial_C G_{AB}(q)\, \partial_\pm q^A \partial_\pm q^B +$$

$$+ 2\, G_{AB}(q)\, \partial_\pm q^A \partial_\pm \xi^B + G_{AB}(q)\, \partial_\pm \eta^A \partial_\pm \eta^B \simeq 0$$

See Ref. 12) where we have applied our method to the case in which the exact solution q_A describes the centre-of-mass motion, i.e., it fulfils

$$m^2 = G_{AB}(q)\, \dot{q}^A(\tau)\, \dot{q}^B(\tau)$$

This defines the $(\text{mass})^2$ of the string. We have computed the mass spectrum and vertex operator in the de Sitter space. The lower mass states are the same as in flat space up to corrections of order g^2 (here $g = \sqrt{\pi\alpha'}/R$), whereas heavy states deviate significantly from the linear Regge trajectories. We find a maximum (very large) value of order $1/g^2$ for the quantum number N and spin J of particles. Application of this method to general cosmological models and to the Schwarzschild geometry will be reported elsewhere. |13|

The critical dimension for bosonic strings is found to be 25 in de Sitter space time and 9 for the supersymmetric string.

The results obtained here for the de Sitter metric show that our perturbative method is well suited to investigate the quantum dynamics of strings in curved space time fully taking into account strong curvature effects.

REFERENCES

1) See, for example: N.D. Birrell and P.C.W. Davies, "Quantum Fields in Curved Space", Cambridge University Press, Cambridge, England, 1982, and references therein.

2) N. Sánchez, Phys.Rev. 24D (1981) 2100;
N. Sánchez and B.F. Whiting, Phys.Rev. 34D (1986) 1056.

3) H.J. de Vega and N. Sánchez, CERN Preprint TH. 4681 (1987), to appear in Nucl. Phys. B.

4) P. Goddard, J. Goldstone, C. Rebbi and C.B. Thorn - Nucl.Phys. B56 (1973) 109.

5) N. Sánchez, CERN preprint TH.4733 (1987), Phys.Lett 195B, 160 (1987).

6) N. Sánchez, Nucl.Phys. B266 (1986) 487.

7) M. Martellini and N. Sánchez, CERN preprint TH. 4680 (1987), Phys.Lett.B., 192B, n°3,4 (1987) 361.

8) N. Sánchez, in the Proceedings of the Second Marcel Grossmann Meeting, Ed. R. Ruffini, (North Holland, 1982) pp.501.

9) N. Sánchez and B.F. Whiting, in "Quantum concepts in space and time", Eds. C. Isham and R. Penrose, (Oxford University Press, 1985), pp.319-324.

10) R.C. Myers and M.J. Perry, Ann. Phys. (N.Y.) 172 (1986) 304.

11) R.C. Myers, Phys. Rev. 35D (1987) 455.

12) H.J. de Vega and N. Sánchez, CERN preprint TH.4784 (1987). Phys.Lett 197B (1987) 320.

13) H.J. de Vega and N. Sánchez, "Quantum dynamics of strings in black hole space times" Paris-Meudon preprint January 1988.

LIST OF PARTICIPANTS

M. Batchelor	Department of Pure Mathematics and Mathematical Statistics University of Cambridge, U.K.
G.G.A. Bäuerle	Institute of Theoretical Physics University of Amsterdam
P.J.M. Bongaarts	Lorentz Institute University of Leiden
R. Brussee	Lorentz Institute University of Leiden
P.A. Bryant	Department of Pure Mathematics and Mathematical Statistics University of Cambridge, U.K.
D.G. Caldi	Department of Physics University of Connecticut, U.S.A.
D. Dunnebier	Department of Mathematics Delft University of Technology
H.N. van Eck	Department of Applied Mathematics University of Twente
W.T. van Est	Mathematical Institute University of Amsterdam
R. Fonk	Lorentz Institute University of Leiden
M. Forger	Fakultät für Physik Albert-Ludwigs-Universität, Freiburg, B.R.D.
J.A. van Gelderen	Department of Mathematics Delft University of Technology

J. de Graaf — Department of Mathematics
Eindhoven University of Technology

P.K.H. Gragert — Department of Applied Mathematics
University of Twente

G.F. Helminck — Department of Applied Mathematics
University of Twente

W.W.J. Hulsbergen — Royal Military Academy, Breda

P.D. Jarvis — Department of Physics
University of Tasmania, Australia

E.M. de Jager — Mathematical Institute
University of Amsterdam

P.H.M. Kersten — Faculty of Applied Mathematics
University of Twente

T.H. Koornwinder — Centre for Mathematics and Computer Science,
Amsterdam

F. ten Kroode — Mathematical Institute
University of Amsterdam

H. Lemei — Department of Mathematics
Delft University of Technology

J.J. Lodder — Institute for Plasma Physics
Rijnhuizen, Nieuwegein

R. Martini — Department of Applied Mathematics
University of Twente

N.C. Masterson — Department of Mathematics
Kings College, London, U.K.

S. Nagamachi — Technical College
Tokushima University, Japan

L. van Nooyen — Lorentz Institute
University of Leiden

F.W. Nijhof — Laboratoire de Physique Théorique
Université Pierre et Marie Curie, Paris, France

C.A.M. Peters — Mathematical Institute
University of Leiden

H.G.J. Pijls — Mathematical Institute
University of Amsterdam

G. Post — Department of Applied Mathematics
University of Twente

H. Rijnks — Department of Mathematics
Delft University of Technology

J.W. de Roever — Department of Applied Mathematics
University of Twente

A. Rogers — Department of Mathematics
Kings College, London, U.K.

N. Sanchez — Département d'Astrophysique Fondamentale
Observatoire de Meudon, France

J.K. Scholma — Centre for Mathematics and Computer Science
Amsterdam

H.J.M. Sterk — Mathematical Institute
University of Nijmegen

L.L.M. Smits — Mathematical Institute
University of Antwerpen, Belgium

G.M. Tuynman — Mathematical Institute
University of Amsterdam

J.P.A. van der Wagt — Department of Applied Mathematics
University of Twente

J. Wess	Institut für Theoretische Physik Universität Karlsruhe, B.R.D.
W. Wesselius	Department of Applied Mathematics University of Twente
B. de Wit	Institute of Theoretical Physics University of Utrecht

Lecture Notes in Mathematics

Vol. 1236: Stochastic Partial Differential Equations and Applications. Proceedings, 1985. Edited by G. Da Prato and L. Tubaro. V, 257 pages. 1987.

Vol. 1237: Rational Approximation and its Applications in Mathematics and Physics. Proceedings, 1985. Edited by J. Gilewicz, M. Pindor and W. Siemaszko. XII, 350 pages. 1987.

Vol. 1250: Stochastic Processes – Mathematics and Physics II. Proceedings 1985. Edited by S. Albeverio, Ph. Blanchard and L. Streit. VI, 359 pages. 1987.

Vol. 1251: Differential Geometric Methods in Mathematical Physics. Proceedings, 1985. Edited by P.L. García and A. Pérez-Rendón. VII, 300 pages. 1987.

Vol. 1255: Differential Geometry and Differential Equations. Proceedings, 1985. Edited by C. Gu, M. Berger and R.L. Bryant. XII, 243 pages. 1987.

Vol. 1256: Pseudo-Differential Operators. Proceedings, 1986. Edited by H.O. Cordes, B. Gramsch and H. Widom. X, 479 pages. 1987.

Vol. 1258: J. Weidmann, Spectral Theory of Ordinary Differential Operators. VI, 303 pages. 1987.

Vol. 1260: N.H. Pavel, Nonlinear Evolution Operators and Semigroups. VI, 285 pages. 1987.

Vol. 1263: V.L. Hansen (Ed.), Differential Geometry. Proceedings, 1985. XI, 288 pages. 1987.

Vol. 1265: W. Van Assche, Asymptotics for Orthogonal Polynomials. VI, 201 pages. 1987.

Vol. 1267: J. Lindenstrauss, V.D. Milman (Eds.), Geometrical Aspects of Functional Analysis. Seminar. VII, 212 pages. 1987.

Vol. 1269: M. Shiota, Nash Manifolds. VI, 223 pages. 1987.

Vol. 1270: C. Carasso, P.-A. Raviart, D. Serre (Eds.), Nonlinear Hyperbolic Problems. Proceedings, 1986. XV, 341 pages. 1987.

Vol. 1272: M.S. Livšic, L.L. Waksman, Commuting Nonselfadjoint Operators in Hilbert Space. III, 115 pages. 1987.

Vol. 1273: G.-M. Greuel, G. Trautmann (Eds.), Singularities, Representation of Algebras, and Vector Bundles. Proceedings, 1985. XIV, 383 pages. 1987.

Vol. 1275: C.A. Berenstein (Ed.), Complex Analysis I. Proceedings, 1985–86. XV, 331 pages. 1987.

Vol. 1276: C.A. Berenstein (Ed.), Complex Analysis II. Proceedings, 1985–86. IX, 320 pages. 1987.

Vol. 1277: C.A. Berenstein (Ed.), Complex Analysis III. Proceedings, 1985–86. X, 350 pages. 1987.

Vol. 1283: S. Mardešić, J. Segal (Eds.), Geometric Topology and Shape Theory. Proceedings, 1986. V, 261 pages. 1987.

Vol. 1285: I.W. Knowles, Y. Saitō (Eds.), Differential Equations and Mathematical Physics. Proceedings, 1986. XVI, 499 pages. 1987.

Vol. 1287: E.B. Saff (Ed.), Approximation Theory, Tampa. Proceedings, 1985–1986. V, 228 pages. 1987.

Vol. 1288: Yu. L. Rodin, Generalized Analytic Functions on Riemann Surfaces. V, 128 pages, 1987.

Vol. 1294: M. Queffélec, Substitution Dynamical Systems – Spectral Analysis. XIII, 240 pages. 1987.

Vol. 1299: S. Watanabe, Yu.V. Prokhorov (Eds.), Probability Theory and Mathematical Statistics. Proceedings, 1986. VIII, 589 pages. 1988.

Vol. 1300: G.B. Seligman, Constructions of Lie Algebras and their Modules. VI, 190 pages. 1988.

Vol. 1302: M. Cwikel, J. Peetre, Y. Sagher, H. Wallin (Eds.), Function Spaces and Applications. Proceedings, 1986. VI, 445 pages. 1988.

Vol. 1303: L. Accardi, W. von Waldenfels (Eds.), Quantum Probability and Applications III. Proceedings, 1987. VI, 373 pages. 1988.

Lecture Notes in Physics

Vol. 290: K.T. Hecht, The Vector Coherent State Method and Its Application to Problems of Higher Symmetries. V, 154 pages. 1987.

Vol. 291: J.L. Linsky, R.E. Stencel (Eds.), Cool Stars, Stellar Systems, and the Sun. Proceedings, 1987. XIII, 537 pages. 1987.

Vol. 292: E.-H. Schröter, M. Schüssler (Eds.), Solar and Stellar Physics. Proceedings, 1987. V, 231 pages. 1987.

Vol. 293: Th. Dorfmüller, R. Pecora (Eds.), Rotational Dynamics of Small and Macromolecules. Proceedings, 1986. V, 249 pages. 1987.

Vol. 294: D. Berényi, G. Hock (Eds.), High-Energy Ion-Atom Collisions. Proceedings, 1987. VIII, 540 pages. 1988.

Vol. 295: P. Schmüser, Feynman-Graphen und Eichtheorien für Experimentalphysiker. VI, 217 Seiten. 1988.

Vol. 296: M. Month, S. Turner (Eds.), Frontiers of Particle Beams. XII, 700 pages. 1988.

Vol. 297: A. Lawrence (Ed.), Comets to Cosmology. X, 415 pages. 1988.

Vol. 298: M. Georgiev, F' Centers in Alkali Halides. XI, 282 pages. 1988.

Vol. 299: J.D. Buckmaster, T. Takeno (Eds.), Mathematical Modeling in Combustion Science. Proceedings, 1987. VI, 168 pages. 1988.

Vol. 300: B.-G. Englert, Semiclassical Theory of Atoms. VII, 401 pages. 1988.

Vol. 301: G. Ferenczi, F. Beleznay (Eds.), New Developments in Semiconductor Physics. Proceedings, 1987. VI, 302 pages. 1988.

Vol. 302: F. Gieres, Geometry of Supersymmetric Gauge Theories. VIII, 189 pages. 1988.

Vol. 303: P. Breitenlohner, D. Maison, K. Sibold (Eds.), Renormalization of Quantum Field Theories with Non-linear Field Transformations. Proceedings, 1987. VI, 239 pages. 1988.

Vol. 304: R. Prud'homme, Fluides hétérogènes et réactifs: écoulements et transferts. VIII, 239 pages. 1988.

Vol. 305: K. Nomoto (Ed.), Atmospheric Diagnostics of Stellar Evolution: Chemical Peculiarity, Mass Loss, and Explosion. Proceedings, 1987. XIV, 468 pages. 1988.

Vol. 306: L. Blitz, F.J. Lockman (Eds.), The Outer Galaxy. Proceedings, 1987. IX, 291 pages. 1988.

Vol. 307: H.R. Miller, P.J. Wiita (Eds.), Active Galactic Nuclei. Proceedings, 1987, XI, 438 pages. 1988.

Vol. 308: H. Bacry, Localizability and Space in Quantum Physics. VII, 81 pages. 1988.

Vol. 309: P.E. Wagner, G. Vali (Eds.), Atmospheric Aerosols and Nucleation. Proceedings, 1988. XVIII, 729 pages. 1988.

Vol. 310: W.C. Seitter, H.W. Duerbeck, M. Tacke (Eds.), Large-Scale Structures in the Universe – Observational and Analytical Methods. Proceedings, 1987. II, 335 pages. 1988.

Vol. 311: P.J.M. Bongaarts, R. Martini (Eds.), Complex Differential Geometry and Supermanifolds in Strings and Fields. Proceedings, 1987. V, 252 pages. 1988.

GPSR Compliance

The European Union's (EU) General Product Safety Regulation (GPSR) is a set of rules that requires consumer products to be safe and our obligations to ensure this.

If you have any concerns about our products, you can contact us on ProductSafety@springernature.com

In case Publisher is established outside the EU, the EU authorized representative is:

Springer Nature Customer Service Center GmbH
Europaplatz 3
69115 Heidelberg, Germany

Batch number: 10620881

Printed by Printforce, the Netherlands